John Brocklesby

Elements of meteorology

With questions for examination, designed for schools and academies

John Brocklesby

Elements of meteorology

With questions for examination, designed for schools and academies

ISBN/EAN: 9783337275365

Printed in Europe, USA, Canada, Australia, Japan

Cover: Foto ©Paul-Georg Meister /pixelio.de

More available books at **www.hansebooks.com**

ELEMENTS
OF
METEOROLOGY,

WITH QUESTIONS FOR EXAMINATION,
DESIGNED FOR SCHOOLS AND ACADEMIES.

BY JOHN BROCKLESBY, A.M.,
Professor of Mathematics and Natural Philosophy in Trinity College, Hartford.

See page 151.

ILLUSTRATED WITH ENGRAVINGS.

TENTH REVISED AND ENLARGED EDITION.

"Fire and hail; snow and vapor;
Stormy wind fulfilling His word."

NEW YORK:
SHELDON AND COMPANY, PUBLISHERS,
498 BROADWAY.
1869.

CASE, LOCKWOOD & CO.,
ELECTROTYPERS AND PRINTERS,
HARTFORD, CONN.

PREFACE.

METEOROLOGY is a subject of interest to all. We live in the very midst of its phenomena, and are constantly subjected to their influence. Many of the singular processes of nature which this science unfolds, are intimately connected with our being and happiness, while others, on account of their beauty and sublimity, fill the mind with admiration and awe.

The subject being one of universal interest, we might naturally suppose it to be universally understood; but such is not the case. Meteorology, as a science, is of recent origin; for it is only within the space of a very few years that it has risen, through the efforts of many gifted minds, to the rank it deserves to hold amid the various departments of knowledge.

Meteorology is a portion of Natural Philosophy, and in the colleges of our land, lectures upon this subject form a part of the regular academical course; but no similar system prevails in our High Schools and Academies. Nor is it to be expected; since, with the present want of facilities for obtaining information, the teacher would be obliged to devote an undue share of his time to the acquisition of the knowledge requisite for this object. Neither can a *text-book* be procured; for the author is not aware that any *distinct treatise* on this science is extant in the English language, except the translation from the German of Kaemtz's "Complete Course of Meteorology;" a work which, though exceedingly valuable to the advanced student, is not suitable for a text-book on account of its *size, expense,* and *mode of execution.*

The present little work has therefore been prepared, not with the view of adding *one more* to the long list of studies now pursued in our academical institutions; but for the purpose of bringing into general notice a rich but hitherto comparatively unknown field, within the domains of natural science.

The author has therefore endeavored, while retaining all the important principles of Meteorology, to condense the subject as much as possible, in order that this elementary treatise may be studied in connection with Natural Philosophy, without consuming too much time.

In regard to *facts,* they have been sought wherever it was supposed they could be found, and reference has been made in nearly all cases to the authorities whence they were taken.

Should it be required a more extended treatise may be expected, adapted to the wants of students in colleges.

HARTFORD, July 7th, 1848.

REFERENCES.

(C. 957), for example, denotes Comstock's Philosophy, Article 957, (last edition.

(Art. 132), for example, denotes Article 132 of this work.

TABLE OF CONTENTS.

	Page
Preface	5
Subject defined	13

PART I.
OF THE ATMOSPHERE.

Barometer	14
Temperature	18
Capillarity	18
Pressure of the Atmosphere	19
Variations in Latitude	19
Variations in Altitude	21
Density of the Atmosphere	22
Weight of the Atmosphere	24
Temperature of the Atmosphere	24
Thermometer	25
Self-registering Thermometer	27
Mean Daily Temperature	29
Variations of Temperature in Latitude	30
Variations of Temperature in Altitude	30
Humidity of the Atmosphere	32
Absolute Humidity	33
Relative Humidity	33
Hygrometer	34
Height of the Atmosphere	36

PART II.
AERIAL PHENOMENA.
CHAPTER I.
OF WINDS IN GENERAL.

Cause of Wind	38
Velocity	39
Anemometer	40
Force of Winds	40
Trade Winds	41
Origin	41
Limits of the Trade Winds	43
Calms	44
Winds of the Higher Latitudes	45
Upper Westerly Wind of the Tropics	48
Periodical Winds	48
Monsoons	48
Origin	49
Land and Sea Breezes	50
Origin	50
Variable Winds	51
Physical Nature of Winds	52
Puna Winds	52
Simoom	53
Sirocco	54

CHAPTER II.
OF HURRICANES.

	Page
Hurricanes defined	54
Path of the Storm	56
Velocity	56
Diameter	57
Examples	57
Fall of the Barometer	58
Circuit Sailing	60
Axis of the Hurricane	61
Espy's Theory	62

CHAPTER III.
OF TORNADOES OR WHIRLWINDS.

Facts	63
Origin	65
Whirlwinds excited by Fires	65
Results of Centrifugal Action	66
Effects of Expansion	67

CHAPTER IV.
OF WATER-SPOUTS.

Water-spouts defined	68
Dimensions	71
Popular Error	72
Sand Pillars	72
Beneficial Effects of Winds	73

PART III.
AQUEOUS PHENOMENA.
CHAPTER I.
OF RAIN.

Cause of Rain	74
Rain Gauge	75
Distribution of Rain in Latitude	75
Exceptions	76
Days of Rain	77
Distribution in Altitude	77
Rain upon Coasts	78
Rains within the Tropics	79
Rainy Seasons	79
Cause	80
Periodical Rains of India	80
Periodical Rains of Congo	81
Rains in the Higher Latitudes	82
Rainy Winds	82
Regions without Rain	83
Egypt	83
Peru	84
Constant Rains	84
Excessive Showers	85
Rain without Clouds	85
Cause	86

CHAPTER II.
OF FOGS.

	Page
Fogs defined	86
Constitution	87
Distribution in Latitude	87
Tropical Regions	87
Temperate Regions	87
Polar Regions	87
Cause	88
Local Distribution	88
Rivers	89
Mountains	90
Capes	91
Shoals	91
Newfoundland	91
England	91
Garuas	92

CHAPTER III.
OF CLOUDS.

Clouds defined	94
Strata of Clouds	95
Thickness	96
Height	96
Clouds on Mountains	97
Classification	99
Cirrus	99
Cumulus	100
Stratus	102
Cirro-stratus	102
Cirro-cumulus	103
Cumulo-stratus	104
Nimbus	105

CHAPTER IV.
OF DEW.

Dew defined	106
Deposition	106
Influence of Condition of the Atmosphere	107
Humidity	107
Serenity	107
Tranquillity	108
Evening and Morning	108
Influence of the Substance bedewed	109
Constitution	109
Surface and Form	109
Location	110
Color	111
Observations	111
Facts Explained	112
Beneficent Distribution	112

CHAPTER V.
OF HOAR-FROST AND SNOW.

Hoar-frost	113
Snow	115

CONTENTS.

	Page
Snow Flake	115
Snow Crystals	116
Natural Snow Balls	118
Red Snow	119
Green Snow	120
Cause	120
Uses of Snow	121

CHAPTER VI.
OF HAIL.

Hail	122
Structure	122
Size	123
Geographical Distribution	123
Origin	123
Volta's Theory	124
Olmsted's Theory	125
Curve of Perpetual Congelation	125
Action of Opposite Currents	126
Action of Whirlwinds	128
Influence of High Mountains	129
Hail in Southern India	129

PART IV.
ELECTRICAL PHENOMENA.
CHAPTER I.
OF ATMOSPHERIC ELECTRICITY.

Electrometers	131
Electric Condition of the Atmosphere	133
Annual Variation in Intensity	133
Daily Variation	133
Variation in Altitude	134
Origin	135
Evaporation	135
Condensation	136
Vegetation	136
Combustion	136
Friction	137

CHAPTER II.
OF THUNDER-STORMS.

General Distribution	137
Origin	138
Electrical State of Thunder-clouds	139
Electric Action of Thunder-clouds	140
Return Stroke	140
Height of Thunder-storms	142
Lightning	143
Origin	143
Kinds	143
Zigzag-lightning	144
Sheet-lightning	144
Ball-lightning	144
Heat-lightning	145
Velocity of Lightning	146
Color	147

CONTENTS. ix

	Page
Effects	147
Fulgurites	148
Volcanic Lightning	149
Thunder	150
Identity of Lightning and Electricity	150
Franklin's Experiment	151
Romas' Experiment	152
Richman's Death	152
Lightning Rod	153
Material	153
Size	153
Mode of Erection	153
Extent of Protection	154
Electric Fogs	155
Spontaneous Electricity	156
St. Elmo's Fire	156
Electric Rain, Hail and Snow	157
Electric Action upon Telegraphic Wires	158

PART V.
OPTICAL PHENOMENA.
CHAPTER I.
OF THE COLOR OF THE ATMOSPHERE AND CLOUDS.

Color of the Atmosphere	162
Cyanometer	162
Effect of Latitude	163
Effect of Altitude	164
Colors of Clouds	165

CHAPTER II.
OF THE RAINBOW.

Primary Bow	170
Secondary Bow	172
Breadth of the Bows	173
Position and Size of the Rainbow	174
Rainbows in the North	175
Extraordinary Bows	175
Supernumerary Bows	176
Lunar Bows	176

CHAPTER III.
OF MIRAGE.

Instances	179
Fata Morgana	181
Origin	183
Erect and Inverted Images above the Object	183
Magnified Images	185
Images below the Object	187
Images produced by Reflection	188
Spectre of the Brocken	190
Artificial Mirage	191

CHAPTER IV.
OF CORONAS AND HALOES.

Coronas	192
Origin	193

	Page
Anthelia	197
Haloes	199
Facts	199
Ordinary Halo of 45°	201
Extraordinary Halo of 90°	203
Circles passing through the Sun	203
Parhelia and Paraselenæ	205

PART VI.
LUMINOUS PHENOMENA.
CHAPTER I.
OF METEORITES.

Facts	206
Size of Meteorites	208
Altitude	208
Velocity	209
Aerolites	209
Form	209
Composition	210
Meteoric Iron	211
Origin of Meteorites	212
First Hypothesis	212
Second Hypothesis	213
Third Hypothesis	213
Fourth Hypothesis	214
Fifth Hypothesis	216

CHAPTER II.
OF SHOOTING-STARS AND METEORIC SHOWERS.

Altitude	216
Velocity	217
Course	218
Magnitude	218
Splendor	219
Meteoric Showers	220
November Epoch	220
Varieties	221
August Epoch	222
Origin	224
Chaldni's Hypothesis	225

CHAPTER III.
OF THE AURORA BOREALIS OR NORTHERN LIGHT.

Constitution	226
Dark Segment	226
Arch of Light	228
Streamers	230
Color	230
Corona	230
Extent	232
Height	233
Sounds attending the Aurora	234
Time	235
Frequency	236
Disturbance of the Magnetic Needle	237
Cause	239
Utility	240

PART VII.
MISCELLANEOUS PHENOMENA.
CHAPTER I.
OF THE FALL OF TERRESTRIAL SUBSTANCES FOREIGN TO THE ATMOSPHERE.

	Page
Dust-storms and Blood-rains	241
Dust-storms	242
Instances	242
Blood-rains	245
Instances	245
Black Rain	247
Red Hail	248
Black Hail	248
Storms of Colored Snow	248
Red Snow	248
Black Snow	249
Nature of the Dust	249
Infusoria	250
Structure	250
The Italian Dust-shower of 1803, and the Calabrian of 1813	251
Atlantic and Cape de Verd Dust	251
Sirocco Dust	253
Number of Distinct Organisms Discovered	256
Number and Extent of Dust-storms and Blood-rains	256
Their Periodicity	257
Origin of the Dust	257
Volcanic Showers	259
Cause	259
Instances—Jorullo	259
Souffriere	259
Tomboro	260
Cosiguina	261
Yellow Rains—Pollen-rains	262
Gossamer-shower	262

CHAPTER II.
DRY FOG AND INDIAN-SUMMER HAZE.

Dry Fog	264
Instances	265
Cause	266
Indian-summer Haze	266
Cause	268

METEOROLOGY.

1. METEOROLOGY. IS THAT BRANCH OF NATURAL SCIENCE WHICH TREATS OF THE ATMOSPHERE AND ITS PHENOMENA. The subject may be properly divided into *six parts.*

2. PART I. THE ATMOSPHERE.

PART II. AERIAL PHENOMENA—*comprehending Winds in general, Hurricanes, Tornadoes, and Water-spouts.*

PART III. AQUEOUS PHENOMENA—*including Rain, Fogs, Clouds, Dew, Hoar-frost and Snow, and Hail.*

PART IV. ELECTRICAL PHENOMENA—*comprising Atmospheric Electricity and Thunder-storms.*

PART V. OPTICAL PHENOMENA—*including the Color of the Atmosphere and Clouds, Rainbow, Mirage, Coronas, and Haloes.*

PART VI. LUMINOUS PHENOMENA—*embracing Meteorites, Shooting Stars and Meteoric Showers, and the Aurora Borealis.*

PART VII. MISCELLANEOUS PHENOMENA—*including the Fall of Terrestrial Substances foreign to the Atmosphere, and Dry Fog and Indian Summer Haze.*

What is Meteorology?
Into how many parts is it divided?
Rehearse the several parts with their subjects.

PART I.

OF THE ATMOSPHERE.

3. As the common properties of the air, viz., *weight*, *fluidity* and *elasticity*, are supposed to be already known, C. 502,) we shall proceed at once to the discussion of the entire body of air, termed the atmosphere; and first of its pressure, which is ascertained by the *barometer*, an instrument so called from the Greek words, *baros*. weight, and *metron*, measure.

BAROMETER.

4. This instrument is of the highest importance in Meteorology, and requires a minute description. It is thus constructed. Into a glass tube, about three feet in length, open at one end and closed at the other, mercury is poured until it is full; the open end being no closed by the finger, or any other means, the tube is i verted, and the lower end immersed in a vessel of me cury. When beneath the surface of the fluid the end is unstopped, and the column of mercury within the tube then settles down, until its summit is about *thirty* inches above the level of that within the vessel. The space above the column in the tube is a void, and is called the *Torricellian vacuum*, from Torricelli, the name of the Italian philosopher, who first constructed this instrument.

5. *The column of mercury within the tube is supported above the level of that in the vessel, by the upward pressure of a column of the atmosphere, having the same base as itself.*

What is the atmosphere?
How is its pressure ascertained?
How is the barometer made?
What supports the column of mercury?

BAROMETER. 15

6. Thus, in fig. 1., the atmospheric column *a a*, of indefinite length, but of the same size as the barometric column Db, presses upon the mercury in the vessel. K, with a force equal to its own weight; now since any force, acting upon a fluid, is communicated in every direction, this pressure will be transmitted through the mercury, in the direction of the arrows, and acts at D, within the tube, against the mercurial column Db. This upward force will be resisted at D, by the weight of Db, and the mercury will sink in the tube until the two pressures counterpoise each other, in exactly the same manner as two equal weights in the opposite scales of a balance.

7. From these considerations, it is manifest, that the *weight of the atmospheric column a a is equal to that of the mercurial column*, Db *of the same base ;* and this weight can be estimated in the following manner. If the base at D contains one square inch, the column Db, at its usual height, will contain, nearly, 30 cubic inches ; and since one cubic inch of mercury weighs 3426.76 grains, the weight of thirty will amount to 102802.8 grains.

This product being now divided by 7004, the number of grains in a pound avoirdupois, the result will be nearly 14.7 lbs. ; a quantity equal to the weight of the barometric column, and consequently to the pressure of the atmosphere on every square inch of surface.

8. Any *increase* in the density of the atmosphere will be denoted by an *elevation* of the mercury, and a *decrease* by its *depression*. The cause of this is obvious, in the first case, *a a* becomes heavier, and requires more

---

Explain Figure 1.
How is the pressure of the atmosphere, on every square inch, computed ?
How does any change in the density of the air affect the height of the barometer ?

mercury to balance it; therefore Db is lengthened. In the second case, *a a* is lighter, and, as a less quantity of mercury will then balance it, Db is shortened. Such changes are constantly occurring, but are very minute, and, in order that they may be accurately indicated, the instrument must be made with the nicest care.

9. To secure a perfect instrument, it is essential that the mercury should be free from any solid impurities, else the summit of the column will either be above, or below, its proper level, according as the foreign matter, mixed with the mercury, is lighter, or heavier, than the fluid. This end is attained by straining the mercury through chamois leather. If it is amalgamated with zinc, or lead, it is purified by washing it with acetic, or sulphuric acid.

10. When the tube is filled, moisture and small bubbles of air are found adhering to its interior surface, and are also contained in the mercury. These, if not expelled, will ascend when the tube is inverted into the Torricellian vacuum, the moisture rising in vapor. By their united elastic force, the ascent of the barometric column will be checked, whenever any increase in the density of the atmosphere tends to elevate it.

11. This source of error is removed by *boiling* the mercury in the tube. When all the air and vapor are expelled, the tube, if gently struck, will give forth a dry, metallic sound; but if a bubble of air remains, the sound will be dull and heavy. By connecting the open end of the tube with an air pump, during the process of boiling, Dr. Jackson, of Boston, has still more effectually removed this imperfection.

12. By these means, the air may perhaps be totally excluded, when the instrument is first constructed; but in the course of time, it will again insinuate itself between the glass and the mercurial column. To prevent this evil, Prof. Daniell, of King's College, London,

---

What precautions are adopted to secure a perfect barometer?
How is the mercury purified, and why?
How are moisture and air expelled from the tube, and why?
What is Prof. Daniell's improvement?

welds to the open end of the glass tube a ring of platinum, which possesses a greater affinity for mercury than glass. The mercury adheres closely to the platinum, like water, and the passage of air, according to all experiments, appears thus to be effectually prevented.

13. Since the constant changes in the weight of the atmosphere produce corresponding fluctuations in the height of the barometer, a scale is placed near the top of the tube, extending from twenty-seven to thirty-one inches, a space, which includes, at the surface of the earth, all the fluctuations of the column. This scale is divided into tenths of an inch; but, as the variations of the barometer are exceedingly minute, a contrivance, called a *vernier*, is annexed, by which a change, to the extent of one five hundredth of an inch, can be easily measured.

14. As the surface of the mercury, in the reservoir, is raised by the descent of the column, and depressed by its elevation, any change in the height of the barometer cannot be accurately estimated, while the scale remains in the same position; unless this surface is always brought to the same point, before taking an observation. The necessity of so doing will be obvious, from the following illustration.

Suppose the surface of the mercury in the cistern K, figure 1., to be *fifty* square inches, while that of a horizontal section of the column is but *one*. Should the barometer sink one inch, the surface of the mercury in the cistern will rise one fiftieth of an inch, and the amount of the depression of the column, if measured from this surface, will be only forty-nine fiftieths of an inch, instead of one inch, its true depression.

15. The contrivance employed by Fortin, a celebrated French artist, to remove this error, consists in adjusting to the cistern K, fig. 1., a movable bottom, which can be elevated or depressed, by means of the screw

---

What is the length of the barometric scale?
How small a variation in height can be measured?
What is Fortin's contrivance, and for what purpose adopted?

P, until the surface of the mercury shall just touch the fixed ivory index L, at its lower extremity; which point is the zero of the scale, or the place from which the height of the barometer begins to be reckoned.

16. When, by adopting the previous precautions, the barometer has been so far perfected, two corrections are still necessary, before recording observations; the first for *temperature*, and the second for *capillarity*. That of temperature depends upon the expansion of the mercury and the scale; the latter being partially corrective of the former, inasmuch as the divisions of measurement upon the scale, lengthen, at the same time, with the column of mercury.

17. TEMPERATURE. Mercury dilates, for every degree Fah. about one ten-thousandth part of its bulk, taken at the freezing point. The expansion of the scale varies with the metal of which it is composed, but its amount is, usually, so small, that it may safely be neglected in the required correction. Hence the following practical rule has been adopted, for reducing any observed altitude of the barometer, to the corresponding altitude, at the freezing point. "*Subtract the ten-thousandth part of the observed altitude, for every degree above the freezing point.*" Thus, if the barometer stands at 29 inches, and the thermometer at 52°, the required correction is $20 \times .0001 \times 29 = 058$. If the temperature is below 32°, the correction must be added. To facilitate these calculations, a thermometer is always attached to the barometer.

18. CAPILLARITY. By *capillary attraction is* (C. 53,) *understood, the force exerted by the interior surface of small tubes, upon the fluids contained within them.* When the fluid moistens the tube, it rises above its proper level; but when it does not, as in the case of mercury, it sinks below it. From this cause, a *depression*, termed its capillarity, occurs in the barometer, the extent

---

How is the barometer affected by a change in temperature?
Give the rule for reducing the height to the corresponding height at the freezing point.
Why is capillarity a source of error?

of which is dependent upon the size of the interior diameter of the tube, and a correction for this must be added to the apparent height, in order to obtain the true altitude. In tubes of a small bore, the error from this source is considerable; but when the diameter exceeds half an inch, it becomes so small, that it may safely be neglected. This will be rendered evident by the inspection of the following table, which gives the amount of depression for tubes of various sizes.

Diameter of tube. inches.	Depression. inches.
.10	.1403
.20	.0581
.40	.0153
.50	.0083

19. When the instrument is not stationary, but is carried from clime to clime, and to different heights above the sea-level, *two* other corrections are necessary; one for the *varying force of gravity*, in different latitudes, and the other for the *change of pressure*, which diminishes with every increase of altitude above the ocean.

Such is the barometer, an instrument of great practical use, and of the highest value in meteorological researches.

PRESSURE OF THE ATMOSPHERE.

20. VARIATION IN LATITUDE. The *mean* or *average* pressure of the atmosphere, as indicated by the barometer, is found to be nearly the same in all latitudes, when every essential correction is made. It increases a little from the equator to about the 30th degree of latitude, where it is *greatest*; it then decreases to nearly the 64th degree, where it is *least*; after this it again increases, and between the 75th and 76th degrees, the pressure is equal to that of the equatorial climes. All

---

Is it greater in tubes of a small or large bore?
When the barometer is portable, what other corrections are necessary?
What is said of the barometer?
In what manner does the pressure of the atmosphere vary in latitude?

this is obvious from the following table, founded upon observations, where corrections are made for gravity, altitude above the sea-level, and temperature.

PLACES.	LATITUDE.	HEIGHT OF BAROMETER.
		Inches.
Cape of Good Hope,	33° S.	29.955
Christianburg,	5° 30′ N.	29.796
Tripoli,	33° N.	30.127
Godthaab,	64° N.	29.593
Spitzbergen,	75° 30′ N.	29.801

21. The pressure of the atmosphere at any given spot is not invariable; for the height of the barometer is perpetually changing throughout the year. The extent of its fluctuations is, however, by no means the same in all places, being *least* at the *equator*, and *greatest* towards the *poles*. Thus its range within the tropics is but a little more than one-fourth of an inch; at New York, 40° 42′ 40″ N. lat., 2.265 inches, from the observations of five years; at St. Johns, Newfoundland, 47° 34′ 3″ N. lat., 2.54 inches, during the same period; while in Great Britain it amounts to *three* inches. The greatest fluctuations occur between the 30th and 60th degrees of latitude.

22. There is also a constant *daily* variation in the atmospheric pressure, for the barometer, as a general rule, falls from 10 o'clock, A. M. to 4, P. M.; it then rises until 10, P. M., when it again begins to descend, reaching its lowest point at 4, A. M.; from this time it rises, until it once more attains its highest elevation, at 10, A. M. These variations are exceedingly minute, and contrary to the annual range, are *greatest* at the *equator*, and *decrease* with the *latitude;* disappearing about the parallel of 60°.

23. This variation amounts at

Give examples.
Where are the annual fluctuations of the barometer least?
Where greatest?
Give examples.
Describe the diurnal variations. Where greatest? Where least?

PLACES.	LATITUDE.	INCHES.
Rio Janeiro,	22° 54' S.	to .067
Lima,	12° 3' S.	to .10
Calcutta,	23° 35' N.	to .072
St. Petersburg,	59° 56' N.	to .005

In the tropical regions, according to Humboldt, so regular are the diurnal changes, that the barometer indicates true time, within a quarter of an hour.

These daily fluctuations, in the atmospheric pressure, for a long time, perplexed meteorologists, but their cause has, at length, been discovered, by means of the late observations, at the English observatories. They are found to arise from the stated variations in temperature, that occur during the day.

24. VARIATIONS IN ALTITUDE. As we ascend above the surface of the earth, we leave a portion of the atmosphere below us, and are freed from its pressure. This fact is denoted by the fall of the barometer. When De Luc, a French philosopher, ascended to the height of 20,000 feet, his barometer sunk to *twelve* inches. In 1838, the aëronaut Green, rose from Vauxhall gardens, in London, to an elevation of nearly three miles and three quarters; the mercury in the barometer gradually descending, from *thirty* inches to *fourteen* and *seven-tenths*.

25. As a general rule, this depression, near the surface of the earth amounts to *one-tenth of an inch for every eighty-seven feet* in altitude; but where perfect accuracy is required, several corrections must be made. The barometer then becomes, in the hands of skillful observers, an important instrument for determining altitudes, and so exact are its indications, that two inde-

---

Give examples.
What is said by Humboldt of their regularity in the tropics?
How are they caused?
How is the pressure of the air influenced by the altitude?
What instrument indicates the changes of pressure?
In the instances given, how low did the mercury sink?
What is the law of depression?
For what purpose is the barometer sometimes employed?
Give instances.

pendent estimates of the height of Mount Ætna, made by means of this instrument, differ only one foot; that of Capt. Smyth being 10,874 feet, while Sir John Herschel's is 10,873 feet.

DENSITY OF THE ATMOSPHERE.

26. When one portion of the atmosphere is said to be more dense than another, all that is meant is simply this; that a *given volume, or bulk, of the first portion, as one gallon, contains more aërial particles than an equal volume of the second;* thus, if it contains twice as many particles, it is said to be twice as dense.

27. The density of the atmosphere *decreases* with the *altitude.* This result is caused by the diminished pressure of the air, and the decreasing force of gravity. Imagine the atmosphere to be divided into a vast number of thin, concentric strata, which in figure 2, are represented by the spaces between the lines 1–2, 2–3, 3–4, 4–5, &c.

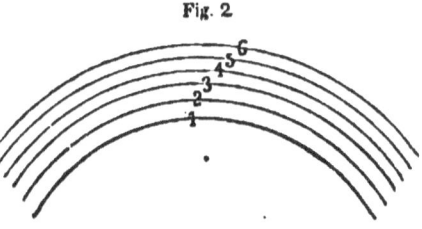

Fig. 2

ATMOSPHERIC STRATA.

Now it is clear, that the particles in each layer *are pressed together* by the whole weight of the atmosphere above them, while, at the same time, they are *drawn together* by the force of gravity. Variations in the latter power are only appreciable at great distances from the earth, and the observed changes in density, at two or more stations, may therefore be ascribed to the difference in the weight or pressure, of the superincumbent atmosphere. The *height* of the barometer, at different elevations, thus denotes the *density* of the air at these points.

When is one portion of the atmosphere denser than another?
What two causes principally influence its density? Describe figure 2.
Which cause may be neglected?
What instrument measures the density?

28. The density, however, is not *exactly* proportioned to the pressure, slight modifications arising, from several causes; the most important of which is *temperature*. The heat of the atmosphere decreases with the altitude, and since heat expands, and cold contracts, a given volume of air, $\frac{1}{480}$th part of its bulk, at 32°, for every degree Fah.; or in other words, thus lessens and increases its density, a correction must be made for this influence.

29. It has been found by calculation, combined with observations, that, if the *altitudes are represented by an increasing arithmetical series, the densities of the atmosphere decrease in a geometrical progression.* Thus, if at the height of 18,000 feet the air, as the barometer indicates, is but half as dense, as at the surface of the earth; at 36,000 feet it will be reduced to one-fourth, and at 54,000 feet to one-eighth of its original density.

30. The rarefaction of the air at lofty elevations, lessens the intensity of sound, impedes respiration, and causes the minute veins of the body to swell and open. Thus, at a short distance, the report of a pistol upon the summit of Mont Blanc, can scarcely be heard. Gay Lussac and Biot, ascending from Paris, in a balloon, to the height of 25,000 feet, breathed with pain and difficulty, and upon the high table lands of Peru, the lips of Dr. Tschudi, cracked and burst; while the blood flowed from the veins of his eyelids.

In consequence of this diminution of pressure, water boils, in such situations, at a comparatively low temperature; thus, at Quito in Equador, 9,537 feet above the sea level, ebullition takes place at the temperature of 196° Fah.

In what manner does temperature affect the density?
What is the law of decrease in reference to altitude? Illustrate.
What are the effects of a rarefied atmosphere? Give instances.

WEIGHT OF THE ATMOSPHERE.

31. We have seen, that a column of mercury, about thirty inches in height, weighs, at the surface of the earth, exactly the same as a column of the atmosphere, possessing the same base. If then the globe was covered with an ocean of mercury, thirty inches in depth, the latter would occupy the identical base that the atmosphere does now, and their respective weights might be regarded as equal.

32. Under this supposition, the diameter of the earth would be increased five feet. The difference then, in cubic feet, between the solidity of the earth, and that of a globe, whose diameter is five feet greater, will equal the number of cubic feet in the sea of mercury. This number multiplied by the weight of a cubic foot of mercury, viz. 848,125 lbs., will equal that of the whole mass, which is the same as the weight of the atmosphere. This calculation has been made, and amounts to more than *five thousand billions of tons.*

TEMPERATURE OF THE ATMOSPHERE.

33. The entire body of air surrounding the globe appears to be warmed in *two* ways; first by the luminous beams of the sun, secondly, by the radiation of heat from the earth.

34. According to Kaemtz and Martin, the atmosphere absorbs nearly one-half the daily amount of heat, which emanates from the sun to the earth, even when the sky is perfectly serene. The remaining portion falling upon the surface of the ground, elevates its temperature, and the earth sends back into the atmosphere rays of invisible heat.

35. Modern researches have shown, that all bodies, through which heat can pass, absorb a greater propor-

---

How is the weight of the atmosphere computed?
How many tons does it weigh?
How is the atmosphere warmed?

tion of non-luminous, than of luminous calorific rays. The heat, therefore, that radiates from the earth, will not pierce the atmosphere, with the power of the solar ray; all will be retained by the lower strata of air, which in their turn, diffuse invisible thermic rays, in every direction.

36. We thus perceive, what all observations have proved, that the upper regions of the atmosphere must be colder than the lower. It is not, however, to be forgotten, that the rarefaction of the superior strata contributes to this condition.

THERMOMETER.

37. The temperature of the atmpsphere is indicated by the *thermometer*, an instrument, which derives its name from the Greek words, *thermos*, warm, and *metron*, measure. It consists of a small glass tube, terminated by a bulb, and is partially filled with mercury.

This fluid is usually preferred for several reasons, the most important of which are, its *uniform dilation*, its *quick susceptibility* to any change in temperature, and the *great range* of its *expansion* in the fluid state. If the instrument is to be exposed to extreme cold, alcohol must be used.

38. As mercury, like other fluids, expands by heat, and contracts by cold, its alternate elevations and depressions within the tube, can be made to indicate the corresponding changes in the state of the air, if *two fixed temperatures can be found, whence to reckon the changes.* These have been discovered. If a thermometer is immersed, at different times, in *melting snow*, the column of mercury invariably sinks to the same place in the tube, though many months may have elapsed between the experiments; and, when exposed to the steam of *boiling water*, the mercury always as-

---

Is it heated most by luminous or non-luminous heat?
Are the *upper* or *lower* regions of the atmosphere the warmest?
How is the temperature of the atmosphere measured?
Describe the thermometer. Why is mercury used?
How are the two fixed temperatures obtained?

cends to the same height, under the same atmospheric pressure.

39. These invariable positions, which are termed the *freezing* and the *boiling* points, are marked upon the scale to which the tube is affixed. In Fahrenheit's thermometer, figure 3., the interval between them is divided into 180 parts, each of which is called a degree (1°) and as the freezing point is marked 32°, the boiling is therefore 212°. The divisions are extended downwards from 32° to 0, or the *zero* point, and when extreme degrees of cold are to be measured, the range is continued to 20°, 40°, and even 60° below zero. If the air is colder than 40° below zero, a spirit thermometer must be used, since mercury becomes solid at this temperature. When Simpson, a late northern traveller, wintered, in 1838, at Fort Confidence, 67° N. lat., he cast a bullet of mercury, the temperature being 49° below zero. Upon firing the ball, it passed through an inch plank, at the distance of ten paces; but flattened and broke against the wall, three or four paces beyond. In addition to the mode of graduation adopted by Fahrenheit, several others prevail (C. 570), which it is not necessary here to discuss.

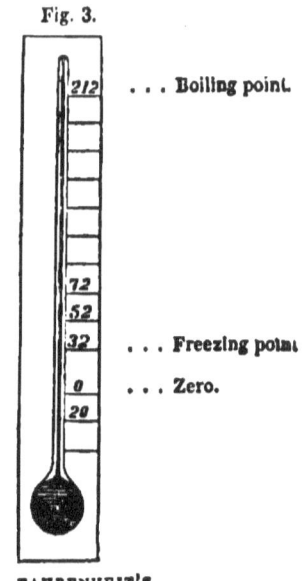

FAHRENHEIT'S THERMOMETER

40. The thermometer employed for meteorological purposes, should be made as accurate as possible, and in

---

Into how many intervals is the space between them divided in Fahrenheit's scale? What are the intervals called?
How many degrees is the freezing point?
How many the boiling point?
What is the zero point?
When must a spirit thermometer be used?
Relate Simpson's experiment.

order to ensure its perfection, many niceties must be observed in its construction.

41. FIRST. *The tube should be of equal size throughout the whole stem;* else the same increase of temperature will not produce the same increase in the height of the mercury, throughout every part of the tube ; and so of the decrease.

SECONDLY. *The bulb should be large in proportion to the tube;* for then slight changes in temperature will be rendered perceptible, and the delicacy of the instrument increased.

THIRDLY. *The mercury should be pure, dry, and recently boiled,* in order to free it from air; and, when in the tube, should there again be boiled, to drive off any air or moisture collected within.

LASTLY. *When the mercury is at the summit of the tube, and every thing else has been expelled, the top of the tube must be perfectly closed by the fusion of the glass, leaving, when the mercury has cooled, a void space or vacuum above.*

42. When a thermometer has been exposed to great changes in temperature during the course of a year, the position of the freezing point upon the scale is found to be somewhat altered; for, if the instrument is then placed in melting snow, the mercury is usually seen to stand a little higher than $32°$, and less than $33°$. This change would occasion a constant error in the observations, and meteorologists therefore verify their thermometers at stated intervals, in the way just mentioned.

SELF-REGISTERING THERMOMETER.

43. The object for which this instrument is constructed, *is to obtain, in the absence of the observer, the highest and lowest temperature of the day, or of any other interval of time.*

What precautions must be taken to construct an accurate thermometer ?
What change occurs in the position of the freezing point ?
How are thermometers verified ?
For what purpose is a self-registering thermometer used ?

One of the most correct thermometers of this kind, now in use, is that invented by Mr. James Six, of Colchester, which is represented in Fig. 4.

It consists of a long glass bulb, G, narrowing into a fine tube, which is first bent downward, forming the arm *a b*, and then upwards, forming the arm *c d*, which terminates in a small cavity, L. The two arms contain mercury, which extends down from *a* on one side, and up to *c* on the other: the bulb and the rest of the tube are filled with alcohol, except the upper part of the cavity L. Upon the top of the mercury in each arm rests an index (which is more perfectly seen at A), consisting of a piece of iron wire capped with enamel, and loosely twined with a fine glass thread; when the mercury descends, the index would fall, were it not for the glass thread, which, pressing like a spring against the sides of the tube, supports the index, in any position.

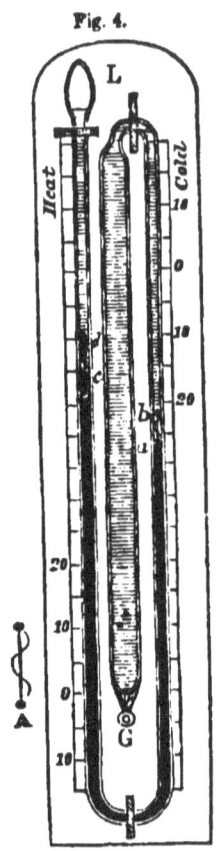

Fig. 4.

SIX'S SELF-REGISTERING THERMOMETER.

44. The action of the instrument is as follows: When an increase of temperature expands the spirit, the mercury is depressed in the arm *a b*, and elevated in *c d*, carrying the index up with it. If the temperature now falls, the spirit contracts, and the mercury descends in *c d*; but the index remains in its last position, from the pressure of the glass spring against the tube; and, as it does not fit tightly to the latter, the alcohol above it flows readily by.

As the cold augments, the mercury rises in *a b*, bearing up the index of this arm, until an increase of temperature occurs, when the mercury here falls, and the

---

Describe Six's, from fig. 4.

index continues stationary. Thus, the *highest point* to which the index rises in the arm, $a\,b$, indicates the *least* temperature, and that in $c\,d$ the *greatest*, that happens in any interval of time, as a day, or a year; and the scale, as is evident from the figure, is graduated accordingly.

45. After every observation, each index requires to be adjusted; this is done by means of a magnet, which, being moved down the side of the arm, draws the index after it.

Another instrument of this kind was invented by Rutherford, (C. 575.)

MEAN DAILY TEMPERATURE.

46. The *mean* or *average* temperature of the day, would be accurately found by observing the thermometer at intervals of an hour during the whole twenty-four, and dividing the sum of the temperatures by the number of observations, viz., twenty-four. This method is however too laborious, and meteorologists have endeavored to arrive at the same result from two or three daily observations.

47. According to Kaemtz, a celebrated German meteorologist, if, in Germany, the thermometer is noted at 6, A. M., 2, P. M., and 10, P. M., and the sum of the temperatures divided by three, the quotient will differ but little from the true mean. The rule adopted in the State of New York, under the direction of the Regents of the University, is as follows:

Mark the temperature, first, between daylight and sunrise; secondly, between 2 and 4, P. M.; thirdly, an hour after sunset: add together the first observation, twice the second and third, and the first of the next day, and divide the sum by six; the result will be the mean.

The mean daily temperature at Philadelphia has been found, from the hourly observations of Capt. Mordecai, to be *one degree less* than the temperature at 9, A. M.

48. By taking the average of all the mean daily temperatures throughout the year, the mean *annual* temperature is ascertained. It is also obtained by the aid of the self-registering thermometer, the average of the two extreme temperatures being regarded as the mean of each day.

49. VARIATIONS OF TEMPERATURE IN LATITUDE. By comparing situations differing widely in latitude, it is found *that the average annual temperature of the atmosphere diminishes from the equator towards either pole.* This will be seen from the annexed table, which presents the results at the sea level, for nine places.

PLACES.	LAT.	TEMP.	PLACES.	LAT.	TEMP.
		Fahren.			Fahren.
Falkland Isles,	51° S.	47° .23	Calcutta, . .	22° 35′ N.	78° .44
Buenos Ayres,	34° 36′ S.	62° .6	Savannah, .	32° 05′ N.	64° .58
Rio Janeiro, .	22° 56′ S.	73° .96	London, . .	51° 31′ N.	50° .72
Maranham, .	2° 29′ S.	81° .32	Melville Isle, .	74° 47′ N.	1.66 below zero.
Trincomalee,	8° 34′ N.	81° .32			

50. From this table it is also evident, that places having the same latitudes, in the two hemispheres, do not necessarily possess the same average temperature. This is owing to a great variety of local causes, the effect of which cannot always be accurately estimated.

51. VARIATIONS IN ALTITUDE. *The temperature of the air diminishes with the altitude,* but the law of decrease is very irregular, being affected by the latitude, seasons, hours of the day, and a diversity of local circumstances. It may however be assumed, as a general rule, *that a loss of heat occurs to the extent of one degree Fah. for every* 343 *feet of elevation.* This is an

---

How is the mean annual temperature found?
How does the temperature of the atmosphere vary in respect to latitude?
Give examples
Do like latitudes in different hemispheres have the same temperature?
How is the temperature affected by altitude?
What is the general law of decrease?

average result, for the rate of decrease is very rapid near the earth, after which it proceeds more slowly, but at the loftiest heights is again accelerated.

52. During the winter of 1838, the French scientific commission stationed at Bossekop, in West Finmark, 69° 58 N. lat., found this law partially reversed, amid the rigors of a polar clime; the temperature of the atmosphere increasing, nearly, 3° Fah. for the first 328 feet in height; beyond this limit it began to decrease, at first slowly, but afterwards with greater rapidity. During the summer, the temperature decreased with the altitude.

53. As a consequence of this gradual reduction of heat, a point at length may be attained, in any latitude, if we continue to ascend, where moisture, once frozen, always remains congealed. Hence, arise the eternal snows and glaciers, that crown the summits of the highest mountains.

54. Since the mean temperature of the air is highest at the equator, and sinks towards either pole, the points of perpetual congelation are farthest removed above the ocean-level within the torrid zone, and gradually approach nearer the general surface of the earth, with the increase of latitude; as the following table shows.

PLACES.	LATITUDE.	LOWER LIMIT OF PERPETUAL SNOW.
Straits of Magellan,	54° S.	3,706 feet.
Chili,	41°	6,009 "
Quito,	00°	15,807 "
Mexico,	19° N.	14,763 "
Ætna,	37° 30'	9,531 "
Kamschatka,	56° 40'	5,249 "
Isle of Mageroe, Norway,	71° 15'	2,362 "

55. A striking departure from the rule exists, however, in India; for while on the south side of the Himmalehs, the snow line occurs at the height of about

---

Was it found true at Bossekop?
What results from this gradual loss of heat?
Where are the points of perpetual congelation nearest to the ocean?
Where farthest from it? Give examples.

13,000 feet, on the northern acclivity it rises to the altitude of 17,000. Many explanations of this singular fact have been given, which admit not of discussion here.

HUMIDITY OF THE ATMOSPHERE.

56. At all temperatures moisture resides in the atmosphere, self-sustained, in an invisible state. Between the particles of air intervals are believed to exist, which are, either partially, or wholly, filled with the vapor that constantly rises from the earth.

57. This peculiarity in the constitution of the atmosphere is termed *the capacity of the air for moisture*, and when the intervals are full of vapor, it is said to be *saturated*. An *increase* of *temperature*, by dilating the air, separates the particles farther from each other; the intervals are thus enlarged, and the capacity of the air *increased*. A *diminution* of *temperature* is followed by contrary effects; the size of the intervals is then reduced, and the capacity *lessened*.

58. The capacity increases, however, at a faster rate than the temperature. A volume of air, at 32° Fah. is capable of containing a quantity of moisture, equal to the 160th part of its own weight; but for every *twenty seven* additional degrees of heat, this quantity is doubled.

Thus a body of air can contain,

At 32° Fah. the 160th part of its own weight.
" 59° " 80th " "
" 86° " 40th " "
" 113° " 20th " "

From this it follows, *that while the temperature advances in an arithmetical series, the capacity is accelerated in a geometrical progression.*

What departure from this rule exists?
What does the atmosphere contain at all temperatures?
What is meant by the capacity of the air for moisture?
When is the air said to be saturated?
What is the effect of heat upon the capacity?
What is the effect of cold?
Which increases at the fastest rate, temperature or capacity?
Give instances. What is the rule in respect to temperature and capacity?

59. ABSOLUTE HUMIDITY. From the cause just mentioned, it would naturally be inferred, that the quantity of atmospheric vapor, or the *absolute humidity*, is greatest in the equinoctial regions, and diminishes towards either pole; a conclusion abundantly supported by facts as will be shown hereafter.

60. The air over the ocean is always saturated, and upon the coasts, in equal latitudes, contains the greatest possible amount of vapor; but the quantity decreases as we advance inland, for the atmosphere of the plains of Oronoco, the steppes of Siberia, and the interior of New Holland, is naturally dry.

61. The absolute humidity diminishes with the altitude, but the rate of reduction is not fully known. By comparing different seasons and hours, it is found to be greater in summer than in winter, and less in the morning than at about mid-day.

62. RELATIVE HUMIDITY. This must not be confounded with absolute humidity. *By relative humidity is understood the dampness of the atmosphere, or its proximity to saturation;* a state dependent upon the mutual influence of its *absolute humidity* and *temperature;* for a given volume of air may be made to pass from a state of dampness, to one of extreme dryness, by merely elevating its temperature, without altering, in the least, the amount of moisture it contains.

Thus one hundred and sixty grains of air, containing one grain of vapor, would be damp at 36° Fah., but hot and withering at the temperature of 90°. By the reverse of this process, a body of hot air will not only become humid, but will even part with a portion of its original moisture, if it is cooled down to any great extent.

63. From the numerous observations of Kaemtz, at Halle, and on the shores of the Baltic, it appears that

---

What is *absolute* humidity? Where is absolute humidity the greatest?
How does it diminish? Where is the air always saturated?
What is said of inland regions? What is the effect of altitude?
Compare summer and winter, morning and mid-day.
What is *relative* humidity? Upon what does it depend? Illustrate the effects of a change of temperature, the absolute humidity being the same.

the relative humidity, in those situations, is highest in the morning before sunrise, and lowest, or farthest removed from the point of saturation, at the hour of the greatest diurnal heat. Corresponding results have been obtained in this country.

HYGROMETER.

64. Those instruments by which the humidity of the atmosphere is measured are called *hygrometers*, from the Greek words *ugros*, moist, and *metron*, measure. Of these there exists a great variety, differing both in form and principle; but those are esteemed the most accurate in their indications, that are constructed upon the principle of condensation, to which allusion has already been made, (Art. 62,) but a more extended explanation is here required.

65. Imagine a brightly polished metallic vessel, partially filled with water, at the temperature of $60°$ Fah., to be placed in a room at the same temperature. If pieces of ice are now thrown into the vessel, the water is gradually cooled down, and as this reduction proceeds, the lustre of the exterior surface will be dimmed, at a certain moment, by a fine *dew*. This is caused by the deposition of moisture from the atmosphere, which, in contact with the cold surface of the vessel, is now cooled down just beyond the point of saturation. The temperature of the water at this instant, which is the same as that of the vessel, is termed *the dew-point*.

66. By marking the difference, in degrees, between the temperature of the air and the dew-point, the relative dryness of the atmosphere, or its remoteness from saturation is obtained. But observations, like these, lead also to other important results; for, by the aid of tables, giving the elastic force of aqueous vapor, at different temperatures, the absolute weight of the vapor, diffused through a given volume of air can be determin-

---

What did Kaemtz observe in respect to relative humidity?
What is a hygrometer? Explain the principle of condensation.
What is the dew-point? How is the relative humidity obtained?
What other results can be deduced?

ed, and likewise the proportion it contains, to that which would be required to saturate it.

67. The hygrometer of Prof. Daniell, which is extensively used, is thus constructed.

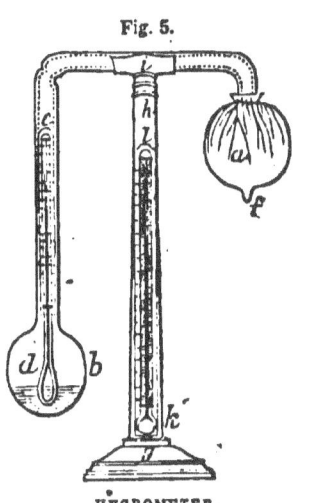

Fig. 5.

HYGROMETER.

A glass tube, *e i*, figure 5., is bent twice at right angles, and terminated by two bulbs, *b* and *f*, of the same material. The bulb *b* is partly filled with ether, into which is inserted the ball of a delicate thermometer, *d*, enclosed in one arm of the instrument. All air is excluded from the tube, which is filled with the vapor of ether; the other bulb, *f*, is covered with a piece of fine muslin, *a*, and upon the pillar, *g h*, a second thermometer, *k l*, is fixed.

68. Observations are thus made. The instrument being placed by an open window, or out of doors, a few drops of good ether are suffered to fall upon the muslin-covered bulb, which, from the rapid evaporation of the ether, soon becomes cool, condensing the ethereal vapor within. In consequence of this effect, the ether in *b* evaporates, thus causing, not only in the ether, but also in the enclosing bulb, a reduction of temperature, which is measured by the interior thermometer, *e d*.

As the evaporation at *a* proceeds, the temperature of *b* still continues to fall, and, at a certain point, the atmospheric vapor will be seen gathering in a ring of dew upon the glass, and the *difference* in degrees, at this moment, between the external and internal thermometer, denotes the relative dryness of the atmosphere. Thus, if on one day the exterior thermometer stood at 65°, and the enclosed sunk to 50° ere the dew-ring appeared—and on another, the former was at 73°, and the latter had descended to 68° before the glass was dimmed

---

Describe Daniell's hygrometer, fig. 5., and explain the mode of taking observations.

with moisture—in the first instance the dryness of the atmosphere would be indicated by 15°, and in the second by 5.

69. The action of this instrument is almost instantaneous, for the enclosed thermometer begins to fall in two seconds after the ether is dropped. It is usual, where great precision is required, to read off the degrees of the interior thermometer at the moment the dew-ring appears, and also at the moment it vanishes; the average of the two observations being taken as the true dew-point.

70. In England the dew-point is seldom 30° Fah. below the temperature of the air; the greatest difference at Hudson, Ohio, as given by Prof. Loomis, is 36°. In the tropical regions its range is the most extensive; for, in the burning clime of India, the dew-point has sometimes sunk as low as 29°, while the temperature of the atmosphere was 90°—a difference of *sixty-one* degrees.

71. HEIGHT OF THE ATMOSPHERE. Whether the atmosphere is boundless or not, is a question which natural philosophers have been unable to determine. De Luc regards it as unlimited, and imagines the planetary spaces to be filled with a medium so exceedingly attenuated as not to retard the motions of the heavenly orbs. The earth and the various celestial bodies are supposed to condense this subtil fluid around them into an atmosphere, by virtue of their respective attractions.

72. Were this true, the densities of the atmospheres thus formed would differ, on account of the variations in the size and mass of these bodies. It therefore constitutes a strong objection to this hypothesis, that the density of the atmosphere of Jupiter (as shown by the refraction of the light of his satellites, at the period of their eclipses) is not superior to that of our own; although the force of attraction at the surface of this planet is al-

---

How far below the temperature of the air does the dew-point descend in England? In Ohio? In India? Is the height of the atmosphere known? What is De Luc's opinion? What is the objection to this hypothesis?

most three times greater than that of the earth. Moreover, when Venus passes near the sun, she exhibits no atmosphere, according to Wollaston, notwithstanding her size is nearly equal to that of the earth.

73. Those who maintain that the atmosphere is limited, suppose, that at a certain distance from the earth, the expansive energy of its particles is exactly balanced by the force of gravity, and that beyond this point, an infinite void extends. This distance has been computed to be not far from 22,200 *miles from the centre of the globe.*

74. Whichever theory may be adopted, it is certain that the atmosphere extends to very great heights. Dr. Wollaston has shown, by calculation, that the atmosphere, at the altitude of nearly *forty miles*, is still sufficiently dense to reflect the rays of the sun, when this luminary is below the horizon. It is capable of transmitting sound at a loftier elevation, for in 1783, a vast meteoric body exploded at an altitude of more than *fifty miles*, the sound reaching the earth like the report of a cannon. Still farther; if the combustion of meteors is truly assigned to the action of the atmosphere, the existence of the latter, at the distance of *one hundred miles* from the earth, may be regarded as proved.

What do the advocates of a limited atmosphere suppose?
How far is this point from the earth's centre?
At what height does the atmosphere reflect light?
At what altitude transmit sound?
What inference is drawn from the combustion of meteors?

PART II.

AERIAL PHENOMENA.

CHAPTER I.

OF WINDS IN GENERAL.

75. CAUSE OF WIND. *Wind is air in motion*, occurring whenever the repose of the atmosphere is broken, from any cause whatsoever. It is usually the result of a change of temperature, and consequently of density, but the rush of an avalanche, causing a sudden displacement of a vast volume of air, has been known to produce a momentary wind of great violence, along the borders of its path.

76. If two contiguous, upright columns of air, with their bases at the same level, are unequally heated, the colder is the denser, and at its base a current will flow towards the lighter column, (just as the compressed air within a bellows streams out into the rarer atmosphere,) but at the top, to supply this loss, a counter current prevails.

77. This is illustrated by Franklin's simple experiment; if a door is opened, communicating between a warm and cold room, and a lighted taper then placed at the bottom of the doorway, the flame is bent *towards* the warm apartment; but if held at the top, its direction is *reversed*.

78. On account of the unequal distribution of heat

What does part second treat of? What does chapter first treat of?
Define wind. When does it occur?
If two contiguous columns of air are unequally heated, what motion takes place? State Franklin's experiment.

over the surface of the globe, phenomena like these occur in nature, on a widely extended scale ; *for if two neighboring countries are unequally heated, the air above them partakes of their respective temperatures, and there arises at the surface of the earth, a wind blowing from the colder to the warmer region, while at the same time, a directly contrary current prevails in the upper strata of the atmosphere.*

79. VELOCITY. Every gradation exists in the speed of winds, from the mildest zephyr, that scarcely bends the flower, to the most violent hurricane, which prostrates the giant oak, and hurls to the ground the proudest works of man. They have been classed as follows, by Smeaton, according to their rapidity and force.

Velocity of the wind, miles per hour.	Perpendicular force on one square foot in lbs. avoirdupois.	Common appellation of such winds.
1	.005	Hardly perceptible.
4	.079	Gentle wind.
5	.123	
10	.492	Pleasant brisk gale.
15	1.107	
20	1.968	Very brisk.
25	3.075	
30	4.429	High wind.
35	6.027	
40	7.873	Very high.
50	12.300	Storm.
60	17.715	Great Storm.
80	31.490	Hurricane.
100	49.200	Violent Hurricane.

80. The velocity of the upper currents of the atmosphere, is as variable as that of the winds which sweep over the surface of the globe; for the aëronaut, Green, who ascended from Liverpool, in 1839, to the height of 14,000 feet, encountered a current that bore him along at the rate of five miles per hour, but upon descending to the altitude of 12,000 feet, he met with a contrary wind, blowing with a velocity of eighty miles per hour.

How does it explain the origin of winds ?
What is said of the velocity of winds ?
Give the common appellations of winds, with their velocity and force.
What is said of the speed of the upper currents ?
Give examples.

On one occasion, his balloon was carried over the space of ninety-seven miles in fifty-eight minutes.

81. ANEMOMETER. The velocity of the wind is estimated by the *anemometer*, an instrument so called from the Greek words, *anemos*, wind, and *metron*, measure. One of the best is Woltmann's. It consists of nothing more than a small windmill, to which is attached an index, in order to mark the number of revolutions per minute; the number of course increasing with the speed of the wind. Now if the atmosphere is still, and the anemometer is carried against it at the rate, for instance, of ten miles per hour, the number of its revolutions will be exactly the same as if the instrument was stationary, and the vanes revolved by the force of a breeze possessing the same velocity.

82. If then, upon a calm day, the anemometer is taken upon a railroad car, moving, for example, at the speed of twenty miles an hour, and the number of revolutions for half an hour accurately noted, we can obtain, (by dividing this result by 30,) the number of revolutions per minute, corresponding to those of a wind having a velocity of twenty miles per hour. In this manner, a table adapted to the instrument can be constructed for all winds, moving with a greater or less rapidity.

The velocity of the higher aërial currents is ascertained by the speed with which the shadow of a cloud passes over the surface of the earth.

83. FORCE. The force of the wind is obtained, by observing the amount of pressure it exerts upon a given, plane surface, perpendicular to its own directions. If the pressure-plate acts freely upon spiral springs, the *power* of the wind is denoted by the *extent* of their *compression*, and that weight will be a measure of its force, which produces the same effect upon the springs.

This instrument, which is also termed an *anemometer*,

What is an anemometer? Describe Woltmann's, and the mode of computing by it the velocity of the wind.

How do we judge of the speed of the upper currents?

In what manner is the force of the wind estimated?

is constructed in exactly the same manner as a letter weigher, where a weight of half an once compresses the spiral, bringing down the index to a certain division of the scale.

84. If, however, the velocities of the different winds are already known, and the force of one obtained, those of the rest can be found by the following rule, viz. *that their forces are as the squares of their velocities.* For instance, if the power of a gale, possessing the speed of twenty miles an hour, is known to be 1,968 pounds on a square foot, that of a storm with a velocity of fifty miles can thus be ascertained by a simple proportion.

(20×20) (50×50)   lbs.   lbs.
400 is to 2500 as 1,968 is to the answer 12,30.

Should the forces be known, it is obvious that the velocities can be computed by reversing this process.

Winds *may be divided into three classes,* CONSTANT, PERIODICAL, and VARIABLE.

CONSTANT WINDS. TRADE WINDS.

85. The most remarkable instance of the first class, is that vast current, which, in the torrid zone, is ever sweeping around the globe, in a westerly direction; and, from its advantage to commerce, in always affording a steady gale to the bark of the adventurous mariner, is denominated the *trade wind.*

86. So uniform is its motion, that on the voyage from the Canaries to Cumana, on the northern coast of South America, it is scarcely necessary to touch a sail; and with equal facility, the richly laden Spanish galleons were accustomed to cross the Pacific from Acapulco to the Philippine Isles.

87. ORIGIN. The cause of this wind has been thus explained by Halley, an English philosopher. From the vertical position of the sun, the regions near the equator

---

If the velocities are known and one force, what else can be obtained?
Give the rule and the example.
If the forces are known, what can be computed?
Into how many classes are winds divided? Name them.
What is the trade wind? How does it originate?

are intensely heated, while those more remotely situated are less so; the temperature gradually diminishing towards either pole. (Art. 49.) In accordance with the principles just unfolded, (Art. 78,) an upper current will flow from the equator towards the poles, and a cold current at the surface of the earth, from the poles and the higher latitudes, towards the equator. Here the air, becoming rarefied by the heat, rises, and mingling with the upper wind flows back again to the polar climes; thus establishing a perpetual circuit. If then the atmosphere was subject to no other influences, *a north wind* would prevail in the torrid zone, in the *northern hemisphere*, and a *south* in the *southern;* but these directions are modified by the rotation of the earth, in the following manner.

88. Every thing upon the surface of the globe at the equator, is carried towards the east, at the rate of about sixty-nine miles in four minutes; but as we recede to the north or south of this line, the eastern velocity is so diminished, that at the latitude of 60° it is reduced to one-half, and at 83° to less than one-eighth of its original amount.

A wind, therefore, blowing from the high latitudes towards the equinoctial clime, is constantly passing into regions where all terrestrial objects have a greater easterly velocity than itself. They will consequently move against it, and as they are apparently stationary, it will thus acquire a relative *westerly* motion. Just as when a traveler, outstripping the wind that blows at his back, feels a breeze directly in his face.

89. Thus, the polar wind in the *northern hemisphere* is influenced by two forces at the same time, one of which carries it to the *south*, and the other to the *west;* and the course it assumes by their combined action must be according to the laws of compound motion, (C. 249,) some intermediate direction, tending from the *north-east* to the *south-west;* and such is the fact, according to all observations.

What two forces influence the polar wind in the northern hemisphere?
What is the direction of the trade wind in this hemisphere?

In a similar manner, the lower current in the *southern hemisphere*, acquires a direction from the *south-east* to the *north-west*.

The passage of a vessel across a river is an illustration in point. If the vessel is steered before the wind, from east to west, while the stream is flowing from north to south, she will be seen by a spectator on shore sailing from north-east to south-west.

90. In the Atlantic and Pacific, the breadth of the *trades* increases as they flow towards the western shores of these vast oceans, the wind gradually changing to the east, by the mutual action of the two currents.

91. The land is heated by the sun far more intensely than the ocean. This is owing to the fact that the solar rays warm only the surface of the earth, scarcely penetrating an inch in the course of a day, while during the same time they pierce the water to the depth of many fathoms. It has been computed that the beams of the sun communicate daily a *hundred times* more heat to a given extent of ground than to an equal surface of water. On this account, the proximity of highly heated continents produces local variations in the direction of these winds; for the air, being more rarefied over the land, ascends, and to supply its place, the cooler air of the trades sets in from the sea towards these localities.

92. Thus, on the African coast, between Cape Bajador and Cape Verde, a north-west wind prevails within the limits of the north-east trade; and off the coast, from Sierra Leone to the Isle of St. Stephen, the trade wind gradually changes to the south and south-west, veering to the west as it approaches the shore. From the same cause, the south-east trade becomes a south wind along the coasts of Chili and Peru.

93. LIMITS OF THE TRADE WINDS. In the Pacific, the north-east trade wind prevails between the 25th

---

What in the southern hemisphere? Illustrate the subject.
What is said of the breadth of the trades?
Why is the land more intensely heated than the ocean? How does this difference cause a local variation in the direction of the trades?
Give instances of such changes. State the limit of the trade winds.

and 2d degree of north latitude. The extent of the south-east trade is not precisely ascertained, but it probably ranges from the 10th to the 21st degree of south latitude. In the Atlantic, the former is comprised between the 30th and 8th degrees of north latitude, and the latter within the limits of the 3d degree of north and the 28th degree of south latitude.

94. The limits, however, are not stationary, but are dependent upon the season—advancing towards the north during the summer of the northern hemisphere, and receding to the south as the sun withdraws to the southern tropic. Thus, on the west coast of Europe, the north-east trade has been found to extend as far as Madeira, and even to Mafra, in Portugal.

95. CALMS. In the vicinity of the Cape Verde isles, between the 8th and 3d degree of north latitude, is a tract denominated by mariners *the rainy sea*. This region is doomed to continual *calms*, broken up only by terrific storms of thunder and lightning, accompanied by torrents of rain. A suffocating heat prevails, and the torpid atmosphere is disturbed, at intervals, by short and sudden gusts, of little extent and power, which blow from every quarter of the heavens, in the space of an hour—each dying away ere it is succeeded by another. In these latitudes, vessels have sometimes been detained for weeks.

In the Pacific, the region of calms is comprised within the 2d degree of north and south latitude, near Cape Francis and the Galapagos islands—a narrow belt of ocean separating the two trades. Here, likewise, dreadful tempests prevail.

96. According to Humboldt, a similar state of the atmosphere exists, during the months of February and March, on the western coast of Mexico, between the 13th and 15th degrees of north latitude, and 103d and 106th degrees of west longitude. A ship, richly laden with cocoa, was here becalmed for the space of twenty-

---

Are the limits stationary? Upon what do they depend?
Give examples. Where is the rainy sea? Describe it.
Where are the calms in the Pacific? What instance is given?

eight days, when the water failing, the crew were compelled by their sufferings to abandon the vessel and seek the shore, eighty leagues distant, in an open boat.

97. The calms are supposed to be caused in the following manner. The adjacent continents to the east of these stagnant regions being far more intensely heated than the sea, the air over the latter would rush easterly towards the land, were it not arrested by a contrary impulse in the direction of the trade wind. If these opposing forces are at any time equally strong, the atmosphere is motionless, and a dead calm ensues— just as a vessel, in ascending a stream, continues stationary when the power of the wind is exactly balanced by that of the current. When, however, the relative strength of these forces rapidly changes, those short and sudden gusts which have been noticed will arise, as one or the other of these impulses prevails.

98. The presence of a highly heated region is strikingly marked in the case of the rainy sea. To the east lies the great African desert, from whose burning surface a vast volume of hot and rarefied air is perpetually rising.

Another cause must not be forgotten, which applies, more particularly, to the calms near Cape Francis. This tract is directly under the equator, and from its peculiar situation, the upward current of rarefied air is probably here so strong as to neutralize the action of the trade winds.

The limits of the calms vary also with the seasons. Thus, in the Atlantic, the range in August is between 3° 15' and 13° N. Lat., but in February, extends from 1° 15' to 6° N. Lat.

WINDS OF THE HIGHER LATITUDES.

99. The upper equatorial currents, flowing off towards either pole, descend, on their passage, to the earth, and since they carry with them an excess of easterly velocity, will become, upon the principles already

explained, (Art. 89,) *south-westerly winds, in the northern hemisphere, and north-westerly in the southern.*

Such would be the course of these currents if left to themselves; but as they meet on their passage with counteracting winds, and are influenced by a variety of causes, their direction is more or less changed; yet not so much, but that a marked predominance in the frequency of westerly winds exists in both hemispheres.

100. That this is true, in regard to the northern hem-

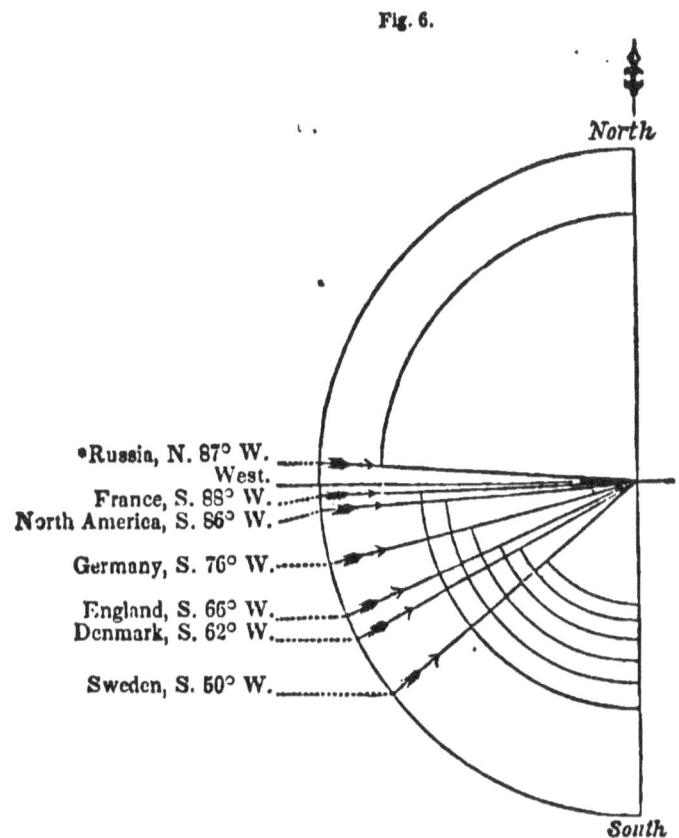

Fig. 6.

PREVAILING DIRECTION OF THE WIND IN DIFFERENT COUNTRIES.

---

* Kaemtz remarks of Russia, "that the number of observations have not been sufficient to enable us to deduce any thing conclusive."

What gives them this direction?
Explain fig. 6., and give the course of the wind for the several countries.

isphere, is obvious from the annexed cut, figure 6., which presents the results of a multitude of observations. A quarter of the circumference of a circle is here supposed to be divided into ninety parts, called *degrees*, and the inclination of the several lines on which the arrows are placed, to the north and south line, measures, in degrees, the average or *mean course of the wind*, in the several places mentioned.

The degrees are reckoned from the south, in all cases except Russia, where they are counted from the north. The points of the arrows indicate the quarter towards which the wind blows.

101. The prevalence of westerly winds in the high latitudes of the north is also shown by the fact, that the average length of the outward passage, by packet, from New York to Liverpool, is but *twenty-three* days, while that of the return voyage is *forty*. It also appears, from the observations of Hamilton, during twenty-six voyages between Philadelphia and Liverpool, extending from 1798 to 1817, that, out of 2029 days on which the wind blew, it came from the west 1101; a result agreeing with the observations of McCord, at Montreal, who found that, from 1836 to 1840, inclusive, the westerly winds at this station constituted more than one-half of all the winds that blew, bearing the ratio of 54 to 100.

102. In the high southern latitudes, the same fact has been observed. Lieut. Maury remarks, that at Cape Horn there are *three times* as many westerly as easterly winds, and that he has seen vessels arrive at Valparaiso and Callao, after having been detained off the Cape, by gales and head winds, for the space of *eighty*, and even *one hundred and twenty days*. In the late Exploring Expedition, the ship Vincennes remained at Orange Harbor, in Terra del Fuego, for the space of sixty days, during which time the weather was exceeding variable; for *thirteen* days the wind blew from the north, eastward, and south-east, while for *forty-seven*, it prevailed from the west.

State facts respecting westerly winds in the high northern latitudes. The same in regard to the high southern latitudes.

UPPER WESTERLY WIND OF THE TROPICS.

103. The prevalence of a westerly wind, above the trades, within the torrid zone, is shown by many conclusive facts. In 1812, ashes from the volcano of St. Vincents were carried easterly, falling upon the island of Barbadoes; and the captain of a Bristol ship declared, that at this time volcanic dust descended to the depth of five inches, upon the deck of his vessel, at the distance of five hundred miles to the east of the former island.

In 1835, an eruption occurred of the volcano of Consanguina, situated in Guatimala. The height of the crater is 3800 feet, and from it issued clouds of ashes, which obscured the sun for five days, and being borne along in a north-easterly direction, by the upper current, fell in the streets of Kingston, Jamaica, seven hundred and thirty miles distant. Even in the latitude of Teneriffe, nearly all travelers have found a westerly wind at the summit of the peak, while the regular trade was blowing in a contrary direction, at the level of the ocean.

PERIODICAL WINDS.

104. MONSOONS. In certain countries within and near the tropics, the regular action of the trade wind is destroyed by the *monsoons*, which are periodical gales, deriving their name from the Malay word *moussin*, signifying seasons. These winds blow, from a certain quarter, for one half of the year, and during the other half from an opposite point; and at the time of their shifting, dead calms, tempests, and variable winds alternately occur.

105. From April to October, the *south-west* monsoon prevails *north* of the equator, and the *south-east* in the *southern* hemisphere; but from October to April, the *north-west* monsoon blows *south* of the equator, and the *north-east* in the *northern* hemisphere.

What is the direction of the wind above the trades? Give the proofs.
What are the monsoons? In what manner do they blow?
From April to October what monsoons prevail, and where?
From October to April what monsoons prevail, and where?

This may be taken as a general rule, subject to the following modification; that the south-west and north-west monsoons occur later in the season, according as the regions over which they prevail are farther removed from the equator.

Thus, in India, at Anjengo, on the Malabar coast, 8° 30′ N. Lat., the south-west monsoon commences as early as the 8th of April; at Bombay, 19° N. Lat., about the 15th of May. In Arabia, it begins a month later than on the shores of Africa, and in the northern part of Ceylon, fifteen or twenty days earlier than on the Coromandel coast.

106. ORIGIN. The cause of these regular changes is to be sought in the effect produced by the sun, during his apparent annual progress from one tropic to the other. In the Indian ocean, for example, as this luminary advances towards the north, the zone of greatest rarefaction recedes from the equator, and the north-east monsoon (which is nothing more than the trade wind) then subsides, and is succeeded by calms and variable winds; but as the summer approaches, and the sun arrives at the northern tropic, the southern portions of the Asiatic continent become hotter than the ocean, and the humid air from the equatorial seas flows northward to the land. South-west winds will therefore arise, (Art. 99,) which prevail from the peninsula of India to the Arabian gulf, until the time of the autumnal equinox. During the same period, the south-east monsoon, in the southern hemisphere, tempers the heat of Lower Guinea, and brings rain to the shores of Brazil.

107. The motions of the atmosphere, however, are *reversed*, as the sun crosses the equator and approaches the southern tropic. Pouring his fervid rays upon Southern Africa, the vast tract of New Holland, and the splendid clime of Brazil, the air flows in from the north and north-west, towards these highly heated regions, and winds from these quarters prevail for several

---

What modifies the general rule? Give examples.
How are the monsoons caused?

months: the monsoon extending along the coast of Brazil, from Cape St. Augustine to the Isle of St. Catherine. But now the influence of the sun is partially withdrawn from Southern Asia; it glows no longer beneath his vertical rays, and over the cooled earth the north-east monsoon resumes its wonted course.

108. LAND AND SEA BREEZES. On the coasts and islands within the tropics a *sea breeze* daily occurs, about nine o'clock in the morning; at first, gently blowing towards the shore, but gradually increasing in force until the middle of the day, when it becomes a brisk gale; after two or three o'clock it begins to subside, and is succeeded at evening by the *land breeze*, which blows freshly off the coast during the night, dying away in the morning, when the sea breeze recommences.

The extent of these winds is variable; in some places they are scarcely noticed beyond the rocks that line the beach; at others they are perceptible three or four leagues from land; while such is their strength on the Malabar coast, that their effects are felt at the distance of *twenty leagues* from shore.

These breezes are occasionally met with in every latitude. They are perceived upon the coasts of the Mediterranean, are sometimes felt at Bergen, in Norway, and even faintly discerned on the shores of Greenland.

109. ORIGIN. During the day, the islands of the tropics acquire a far more elevated temperature than the adjacent ocean; (Art. 91,) the atmosphere above them partakes of their warmth, and currents of rarefied air ascend from the interior of the land. To supply the partial void thus created, the cool, dense air of the ocean flows in from every quarter towards the shore, and the *sea breeze* then prevails.

About mid-day the sea breeze is strongest, since the velocity and force of the ascending current is then at its height, for the sun now acts with its greatest energy;

Describe the land and sea breezes. Where do they prevail?
How do they originate?

but as this luminary descends in the heavens, and sinks beneath the waves, the land rapidly loses its heat by radiation, while the temperature of the ocean at its surface is scarcely changed. This is owing to the fact already stated, that the rays warm only the *surface* of the earth, but are diffused through the water to a considerable depth ; and besides, whenever the upper stratum of the fluid is cooled, it becomes heavier and sinks ; and a warmer stratum rising to the top the surface thus maintains an almost unvarying temperature. For these reasons, the land, at length, becomes colder than the sea, while the air above it, acquiring its temperature, is condensed, and flowing off in every direction to the warm and lighter atmosphere that floats above the ocean, gives rise to the *land breeze*, which prevails throughout the night.

110. VARIABLE WINDS. From the extreme mobility of the air, the direction of the wind is affected by a countless variety of causes, *such as the nature of the soil, the inequalities of its surface, the vicinity of the ocean and of lakes ; and the temperature, course and proximity of mountains.*

These local influences are, for the most part, controlled, where the great aërial currents exist in all their power ; but in the extra-tropical regions, where the force of the latter is diminished, a perpetual contest occurs between the permanent and temporary currents, giving rise to constant fluctuations in the strength and direction of the wind.

111. It appears from observations made at Toronto, and at Hudson, Ohio, that although the wind blows from every point of the compass during the year, yet, such is the force of the northerly gales, that, in these latitudes, there is a general motion of the atmosphere from N. W. to S. E. In England, on the contrary, from the hourly observations made at Plymouth, there seems

---

In what regions do variable winds prevail ? What appears to be the general course of the atmosphere at Toronto and Hudson ?
What in England ?

to be an annual movement of the atmosphere, from the S. S. E. towards the N. N. W.

112. PHYSICAL NATURE OF WINDS. Winds are hot, cold, dry or moist, according to *the direction whence they blow, and the kind of surface over which they pass.* In Europe the westerly winds are warm and moist, and the north-easterly cold and dry; for the former come over the sea from the lower latitudes, and the latter sweep across the land from the polar climes: in our own climate, a north-easterly wind is cold and moist.

A south wind in the northern hemisphere is warm and humid, since it comes from warmer regions, and its capacity for moisture is constantly diminishing in its northward progress; from opposite causes a north wind is keen and dry.

In the southern hemisphere the nature of these winds would be interchanged.

113. PUNA WINDS. In Peru, between the Cordilleras and the Andes, at the height of 12,000 feet, are vast tracts of desolate table-land, known by the name of the *Puna.* These regions are swept, for four months in the year, by a piercingly cold wind from the snowy peaks of the Cordilleras, which is so extremely dry, and absorbs with such rapidity the moisture of animal bodies, that it prevents putridity. If a mule happens to die upon these plains, it is converted, in the course of a few days, into a mummy, even the entrails being free from the slightest evidence of decay.

According to Prescott, the ancient Peruvians preserved the bodies of their dead for ages, by simply exposing them to the dry, cold, and rarefied atmosphere of the mountains.

114. SIMOOM. Upon the arid plains of Asia, and especially on the vast deserts of Africa, an intensely hot

---

Whence arise the differences in the properties of wind?
What is the nature of a south wind in the northern hemisphere, and why? Of a north wind, and why?
Why would their properties be reversed in the southern hemisphere?
Describe the Puna winds. What fact is stated by Prescott?

wind occasionally prevails. In Arabia and Syria, it is known by the name of the *simoom*, from the Arabic word *samma*, signifying at once *hot* and *poisonous*. In Egypt it is termed *chamsin*, *fifty*, because it usually continues *fifty days;* while in the western parts of the great Zahara, along the Senegal, and upon the coast of Guinea, it receives the name of *harmattan*.

The stories of the Arabs, and the accounts of the earlier travelers, in regard to this wind, are clothed with marvelous fictions. It is described as a poisonous, fiery blast, that instantly destroys life; none ever surviving the effects of its deadly influence, if once inhaled. But these fables are now exploded, and the simoom is known to possess no other properties than those which naturally belong to an exceedingly hot and parching wind.

115. CAUSE. Its origin is to be sought in the peculiarities of the soil, and the geographical position of the countries over which it reigns.

The surface of the Asiatic and African deserts is composed of dry quartz sand, which the powerful, vertical rays of the sun render burning to the touch. The heat of these regions is insupportable, and their atmosphere like the breath of a furnace. In June, 1813, at Esné, in Upper Egypt, the thermometer of Burckardt rose to 120° Fah. beneath the roof of a tent, and in 1841, the British embassy to the king of Shoa, while advancing from Tajura to Abyssinia, suffered under a temperature of 126° Fah. in the shade.

When, under such circumstances, the wind rises and sweeps these burning wastes, it is at the same time hot, and extremely destitute of moisture; and, as it bears aloft the fine particles of sand, the atmosphere is tinged with a *reddish*, or *purple haze*, the sure precursor of the simoom.

116. Though the blast of the simoom inflicts not instant death, it is yet a dreadful visitant to the traveler

---

What is the simoom? Where does it prevail?
By what other names is it known? What is the truth respecting it?
How does it originate?

of the desert. Clouds of glowing sand, at times so thick that objects are invisible at the distance of a few paces, are driven with blinding force against the face; the moisture is rapidly absorbed from the body, the skin becomes parched, the throat inflamed, respiration is accelerated, and a raging thirst created. And in the midst of these horrors, the burning blast deprives its unhappy victims of the only means which they possess for alleviating their sufferings; the water evaporates through the skins in which it is carried, and whole caravans have been known to perish, the prey of a consuming thirst.

117. SIROCCO. This name is given to a south-east wind which prevails in the Mediterranean isles, and along the Italian shores. During the summer and autumn it is peculiarly distressing to the inhabitants of these regions; an oppressive sensation of heat is then felt, the skin is bathed in perspiration, the body becomes weak and languid, and the mind dispirited. These effects are attributed to the fact, that the sirocco, at this time, is both *hot* and *moist;* very little evaporation therefore occurs, and the sensations experienced, under these circumstances, are similar to those which are felt during a very sultry state of the atmosphere. While this wind prevails, the air is obscured by fine particles of *dust*, and is always *hazy*.

The sirocco has been generally supposed to arise from a current of air flowing across the Mediterranean from the glowing sands of Africa. It acquires its heat from the desert, and its moisture from the sea.

CHAPTER II.

OF HURRICANES.

118. HURRICANES *are terrific storms, accompanied, at times, by thunder and lightning; and differ from*

Describe its effects. Describe the Sirocco.
What does chapter second treat of? What are hurricanes?

HURRICANES.

every other kind of tempest by their extent, their irresistible power, and the sudden changes that occur in the direction of the wind. Though known in other climes, they rage with the greatest fury in the tropical regions. The rich products of the plantations are destroyed in a moment, forests are leveled, the firmest edifices prostrated and their roofs whirled aloft into the air, which is filled with the flying fragments of a thousand ruins. Upon the coasts, the waves rush landward with appalling violence, lining the harbors and the adjacent shores with the cargoes and wrecks of shattered vessels.

119. From the late independent investigations of several eminent philosophers, it also appears *that hurricanes are extensive storms of wind, which revolve around an axis, either upright or inclined to the horizon; while at the same time, the body of the storm has a progressive motion over the surface of the globe.*

120. We learn from the numerous observations collected by Mr. Redfield, of New York, that in the northern hemisphere, the Atlantic hurricanes generally originate to the east of the Carribean islands, and that their path is from *southeast* to *north-west*, until they have passed the northern tropic, when their course changes from *south-west* to *north-east;* the rotation of the storm being from *right to left, contrary to the motion of the sun,* (see fig. 7., where the arrows show the direction of the wind.)

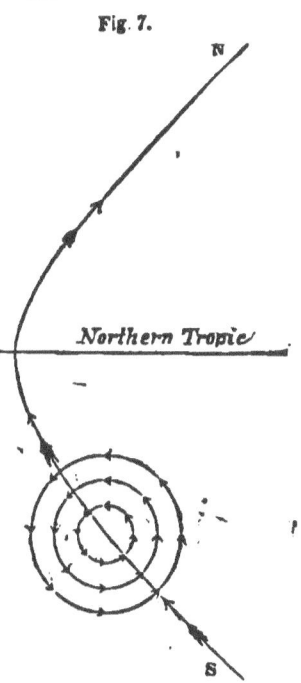

Fig. 7.

GENERAL DIRECTION AND ROTATION OF HURRICANES IN THE NORTHERN HEMISPHERE.

Where are they most violent? What do many philosophers now consider them to be?

State Mr. Redfield's views in regard to the Atlantic hurricanes of the northern hemisphere.

The researches of Col. Reid, the Governor of the Bermudas, have likewise shown, that the storms and tempests of the southern latitudes are *vast whirlwinds;* moving, however, in a different manner from the hurricanes of the northern hemisphere. Thus, *south of the equator*, the general course of the hurricanes is from the *north-east* to the *south-west*, within the southern tropic; but after passing this limit they proceed from the *north-west* to the *south-east;* revolving from *left to right, in the same way as the sun;* a fact previously conjectured by Mr. Redfield. (See fig. 8.)

The hurricanes of the southern hemisphere frequently occur in the vicinity of Mauritius and Madagascar.

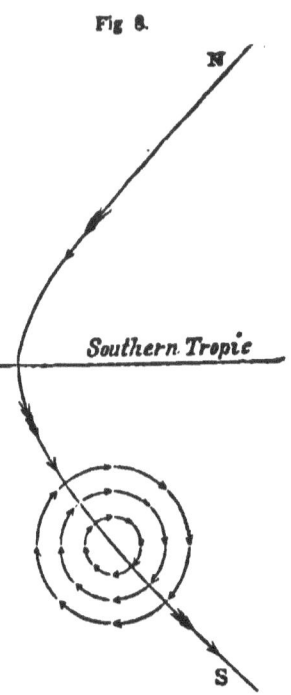

Fig 8.

GENERAL DIRECTION AND ROTATION OF HURRICANES IN THE SOUTHERN HEMISPHERE

121. PATH OF THE STORM. The distance traversed by these desolating tempests is immense. The memorable gale of August, 1830, which fell upon St. Thomas, in the West Indies, on the 12th, reached the Banks of Newfoundland on the 19th; having traveled more than *three thousand nautical miles in seven days;* and the observed track of the Cuba hurricane of 1844 was but little inferior in length.

122. VELOCITY. Their progressive velocity varies on the Atlantic Ocean, from seventeen to thirty miles per hour; but at certain portions of the track it is sometimes much higher; as in the case of the Cuba hurricane, where the

---

State Col. Reid's views in respect to those of the southern.
What is said of the distance traversed by hurricanes?
What of their progressive and rotary velocity?

average rate from the Bahamas to 45° N. Lat. was forty miles per hour. Distinct from the *progressive* is the *rotary* velocity, which increases from the exterior boundary to the centre of the storm, near which point the tempest rages with terrific force ; the wind sometimes blowing at the rate of one hundred miles per hour.

123. DIAMETER. The surface simultaneously swept by these tremendous whirlwinds is a vast circle, varying from one hundred to five hundred miles in diameter ; but even the greatest of these dimensions was exceeded in the Cuba hurricane, for its breadth was computed by Mr. Redfield to be at least 800 miles, and the area over which it prevailed, throughout its whole length, 2,400,000 square miles ; an extent of surface equal to *two-thirds of that of all Europe.*

124. The rotary character of the hurricane accounts for the frequent *changes* that occur in the direction of the wind ; since, in order to preserve a circular motion, there must be a constant deflection from a straight course, and, at corresponding points in each half of the storm, the gale will blow from opposite quarters. The changes thus caused, will be perceived at any spot over which this fearful visitant passes.

It also explains the fact, that the violence of the wind *increases* towards the *centre,* and that, within the very vortex of the hurricane, the air is in repose. Here occurs that awful calm, described by mariners as the *lull of the tempest,* in which it seems to sleep, only to gather strength for mightier conflicts.

125. CASES. Numerous instances of the facts above mentioned might be adduced, but one or two will suffice. In the Antigua hurricane of 1837, described by Col. Reid, it appears that Capt. Newby of the Water Witch, first experienced its effects at St. Thomas, in the West Indies, on the morning of the second of August. The wind was then N. N. W., and at three in the afternoon

---

How great is their breadth ?
How great the surface over which they prevail ?
What facts are explained by the rotation of the storm ?
Give instances.

became violent. At five P. M. it blew a severe gale, and at seven P. M., says Capt. Newby, "a hurricane arose beyond description dreadful. Soon after a calm succeeded for about ten minutes, and then, in the most tremendous screech I ever heard, it recommenced from the S. and S. W. At two o'clock on the morning of the third, the gale somewhat abated, and the *barometer rose an inch*. At daylight, out of forty vessels, the Water Witch and one other were the only two not sunk, ashore, or capsized."

126. On the 12th of August, 1837, another hurricane commenced, in the same region, in 17° N. Lat. and 53° 45' W. Lon. At midnight on the 18th, in 31° N. Lat., the ship Rawlin, Capt. Macqueen, appears, according to Col. Reid, to have been in the very vortex of the storm. On the 17th, the wind blew strong from the N. E. by E. for twelve hours, then suddenly changed to the north, blowing with undiminished violence till the 18th at midnight when, in an instant, a perfect calm ensued for the space of one hour; then, "quick as thought, the hurricane sprung up with tremendous force from the S. W.; no premonitory swell of the wind preceding the convulsion." During the gale, the barometer was almost invisible in the tube above the framework of the instrument.

The sudden and extraordinary transition detailed in the cases just cited, are fully explained by supposing, that the vessels passed from one side of the whirl to the other, through the vortex of the tempest.

127. FALL OF THE BAROMETER. If the hurricane is indeed a vast whirlwind, the atmosphere, constituting the body of the storm, will be driven outward from the centre towards the margin (C. 171), just as water in a pail, which is made to revolve rapidly, flies from the centre, and swells up the sides. But the pressure of the atmosphere, beyond the whirl, checking, and resisting this centrifugal force, at length arrests the outward progress of the aërial particles, and limits the storm.

If the hurricane is a whirlwind, in what manner should the barometer fall and rise?

We should consequently expect to find (in accordance with the laws of circular motion) the density of the air increasing from the centre to the circumference of the storm, and even for some distance beyond its boundary; and likewise, that when a hurricane passed diametrically over any region, the atmospheric pressure would decrease, and the *barometer* continue to *sink*, during the *first half of the storm;* but that the instrument would gradually *rise*, as the *last half* passed over. Such indeed is the case; for, amid all the phenomena of storms, no fact is better established than this, *that an extraordinary depression of the barometer in tropical climates is a sure forerunner of a hurricane.*

128. Before the tempest of Aug. 2d, 1837, the harbor-master of Porto Rico warned the shipping in port to prepare against a storm, as the barometer was falling in an unusual manner; having sunk one and a half inches since 8 o'clock in the evening of the preceding day. All precautions were however in vain; thirty-three vessels at anchor were destroyed, and, at St. Bartholomews, two hundred and fifty buildings levelled to the earth. The following table of observations, taken at St. Thomas, over which island this hurricane passed, is full of instruction in regard to this important point.

Is this the case? Relate the instances given.

129.

TIME.	HEIGHT OF THE BAROMETER.	WIND.
Hours and Minutes.	Inches.	Direction and Force.
Aug. 2d, A. M. 6	29.95	
P. M. 2 10	29.77	N. W. ⎫ Increasing
3 45	29.69	N. ⎬ Tempest.
4 45	29.51	N. ⎭
5 45	29.33	N. E.
6 30	29.16	N. W.
6 35	28.93	N. W.
6 45	28.80	N. W. ⎬ Hurricane.
7 10	28.62	N. W.
7 30	28.18	N. W.
7 35	28.13	
7 52	28.09	⎬ Dead Calm.
8 10	28.09	
8 20	28.09	
8 23	28.44	S. S. E.
8 33	28.53	S. E.
8 38	28.62	S. E.
8 45	28.71	S. E.
8 50	28.80	S. E.
9	28.99	S. E.
9 10	29.16	S. E. ⎬ Hurricane.
9 25	29.24	S. E.
9 35	29.33	S. E.
9 50	29.42	S. E.
10 10	29.51	S. E.
10 35	29.60	S. E.
11 30	29.64	S. E.
Aug. 3d, A. M. 2 45	29.78	S. E.
8	29.91	S. W.
9	29.93	E.

130. In the case of the Water Witch, we have seen, that, when the centre of the tempest was past, and the gale abated, the barometer rose an inch.

131. CIRCUIT SAILING. The gyratory motion of hurricanes is strikingly evinced by vessels sailing on a circular course, when scudding before the wind. The most remarkable case is that of the Charles Heddle, related by Mr. Piddington, which occurred in a storm, near Mauritius, in Feb. 1845.

It appears from the log-book of this ship, that, in her course before the gale the wind changed completely

---

What example is given of circuit sailing?

round five times in the space of one hundred and seventeen hours, having an average velocity of eleven miles and seven-tenths per hour. The *whole distance* thus sailed by the vessel was *thirteen hundred and seventy-three* miles; while her *actual progress* during this time in a south-westerly direction, was found to be *only three hundred and fifty-four miles.*

132. AXIS OF THE HURRICANE. The axis of the hurricane is not, necessarily, upright, but is usually inclined to the horizon; leaning in the direction which the tempest takes. This is owing to the friction of the base of the hurricane against the surface of the earth. Its velocity is thus checked, while the upper portion is driven forward, and overhangs the base.

This position of the axis is indicated by the circumstance that the tokens of the approaching tempest often appear in the higher regions of the atmosphere, before it is felt below. The navigators of the tropic seas sometimes behold, high in the air, a small black cloud; rapidly it spreads down to the horizon, shrouding sea and sky, and the tempest then suddenly descends upon them in all its fury.

133. REMARKS. Such are the opinions entertained by Redfield, Reid, Dove, and others, in regard to storms and hurricanes; opinions based upon a vast assemblage of facts and observations, gathered from all points, within the track of a great number of these desolating gales. The numerous observations taken upon the American coast, commensurate with the extent of the Atlantic tempests, have been systematized by Mr. W. C. Redfield, of New York; while Col. Reid has investigated the West India hurricanes, and those of the southern hemisphere, with great success. The log-books of the British navy, in which the phenomena of the weather are recorded every half hour, have been

---

What is the position of the axis of the hurricane?
How is it caused?
How is this position sometimes indicated?
Detail the labors of Redfield, Reid, and Dove.

placed at his disposal, and he has thus been furnished with an immense collection of valuable facts. Prof. Dove, of Berlin, has studied the laws of hurricanes in Europe, and gathered a large number of observations from every quarter of the globe. By noticing the time and place of each observation, storm-charts have been constructed for the use of mariners, and it is highly in favor of the rotary theory, that the conclusions resulting from these extensive and independent investigations are substantially the same.

134. ESPY's THEORY. The rotary character of hurricanes, including tornadoes and water-spouts, is however denied by Mr. Espy, of Philadelphia, who maintains *that the wind blows from every quarter towards the centre of the storm.* Espy asserts, that this law obtains without a single exception, in seventeen storms which he has investigated. The influx of wind towards the centre, he supposes to be caused by the *development of heat*, which occurs whenever atmospheric vapor is condensed in the form of a cloud. The heat, thus disengaged, rarefies the surrounding air, and establishes an upward current; and so great an expansion is believed, at times, to result from this cause, that the velocity of the ascending current has been computed to exceed three hundred and fifty feet per second.

To this point of greatest rarefaction, the atmosphere rushes in from every side, just as the air of a room flows towards the heated current of the chimney; the violence of the wind depending upon the rate of speed in the ascending column. Most of the phenomena of meteorology are also explained by Mr. Espy in accordance with his peculiar views.

135. The centripetal theory has found many able supporters; but that of Redfield and Reid has been more generally adopted by men of science.

136. It may perhaps be found, when our investigations are multiplied and more extended, that both these

---

Detail Mr. Espy's theory.
Which theory has been more generally adopted?
May these two motions co-exist?

motions often co-exist; a circumstance which is by no means impossible. For when a whirlwind is once in motion, from any cause whatsoever, the great rarefaction of air that occurs at the centre, will create an influx of the atmosphere towards this point from all quarters, except where it is opposed by the centrifugal force. Now if the base of the whirl is above the surface of the earth, or when touching it, is inclined to it, (which is usually the case,) currents of air will flow beneath the base towards the vortex, and evidences of centripetal action will not be wanting.

CHAPTER III.

OF TORNADOES OR WHIRLWINDS.

137. Tornadoes may be regarded as hurricanes, differing chiefly in respect to their *extent* and *continuance*. They last only from fifteen to sixty or seventy seconds, their breadth varies from a few rods to several hundred yards, and it is probable that the length of their track rarely exceeds twenty-five miles.

138. FACTS. This phenomenon is usually preceded by a calm and sultry state of the atmosphere; when suddenly the whirlwind appears, traversing the earth with great velocity, and sweeping down by its tremendous power the mightiest products of nature, and the strongest works of man. Ponderous bodies are whirled aloft into the air; trees of large dimensions twisted off or torn up by the roots; buildings of the firmest construction prostrated, and streams whirled from their beds and their channels laid bare. A whirlwind that occurred in Silesia, in the year 1820, carried a mass weighing more than 650 lbs., *fifty* feet above the top of a house, and

---

What are tornadoes?
Describe their effects.
By what phenomena are they attended?

deposited it on the other side in a ditch, *one hundred and fifty* paces distant.

139. In 1755 a tornado fell upon the village of Mirabeau, in Burgundy, laying dry the channel of the small river by which it is traversed, and carrying the stream to the distance of sixty paces. In the New Haven whirlwind of 1839, and in that which occurred at Chatenay, near Paris, during the same year, trees eighteen inches in diameter were torn up by the roots. In one which happened at Maysville, Ohio, in 1842, a barn containing *three tons* of hay and *four horses*, was lifted entirely from its foundations. And such was the force of the wind during a tornado which occurred at Calcutta in 1833, that a bamboo was driven quite through a wall *five feet thick*, covered with masonry on both sides; an effect which was estimated, by a person on the spot, to be equal to that produced by a cannon carrying a six-pound ball.

By the action of a tornado, fowls are often entirely stripped of their feathers, and light substances carried to a distance varying from two to twenty miles.

140. The whirlwind is attended by all the usual phenomena of thunder-storms; showers of hail frequently occur, and, at times, it is the scene of very extraordinary electric appearances. In the one which happened at Morgan, Ohio, on the night of the 19th of June, 1823, a bright cloud of the *color of a glowing oven*, and apparently half an acre in extent, was seen moving below the dark canopy of the tempest. It shone with a splendor above that of the full moon, and ten minutes after its passage, the narrator of the phenomena was enabled to read his Bible by its light. Just before the Shelbyville tornado, which took place at midnight, on the 31st of May, 1830, two luminous clouds were seen approaching each other, of the *color of red hot iron;* for a moment they united above the town, extending over it like two fiery wings, and, at the next, rushed down to the earth: at this instant the whirlwind burst in all its fury upon the devoted spot. The writer

---

What extraordinary appearances are sometimes seen?

of this work was informed by an eye-witness, that, during the prevalence of the storm, so incessant was the play of the lightning, that the titles of books could be easily read, and the use of lamps was discarded in going to different parts of the house.

141. ORIGIN. Several theories have been advanced to explain the causes of whirlwinds, but they are supposed to be generally produced by the *lateral action of opposing winds*, or the influence of a brisk gale upon a portion of the atmosphere in repose; in a manner analogous to the eddies that arise at the junction of two streams, flowing with unequal velocities, or the airwhirls that occur, when a wind sweeps by the corner of a building, and strikes the calm air beyond it.

142. The existence of such opposing currents is fully proved by the observations of aëronauts, as well as by those of observers at the surface of the globe.

The whirl appears to originate in the higher regions of the atmosphere, and as it increases in violence, to descend; its base gradually approaching until it touches the earth.

Thus, when on the summit of the Rigi—a mountain in Switzerland—Kaemtz beheld two masses of fog approaching each other, in the valley of Goldan, while the air around him was calm, and the sky serene. As soon as they united, a gyratory motion was perceived, the fog rapidly extended, accompanied with violent gusts of rain and hail. At the same time, (as appeared from subsequent information,) a furious storm fell upon the lake of the Four Cantons, far below; in the midst of which a water-spout was seen. (Art. 150.)

143. WHIRLWINDS EXCITED BY FIRES. Extensive conflagrations have been known also to produce whirlwinds, in consequence of the strong upward current, resulting from the great expansion of the heated air.

A remarkable instance of this kind occurred between

How do they originate, and where?
What did Kaemtz witness?
What is the effect of extensive fires?

Great Barrington and Stockbridge, Mass., in the month of April, 1783, and is thus related by Mr. T. Dwight, who beheld it. "In an open field, a large quantity of brush-wood was lying in rows and heaps for burning, perfectly dry and combustible. On a certain day, when the atmosphere was entirely calm, the brush was ignited on all sides of the field at once. I was residing at this time, at the distance of about half a mile from the fire, when suddenly my attention was aroused by a loud, roaring noise, like heavy thunder. Upon going to the door, I beheld the fire covering the field, and the flames collected from every side into a *fiery column*, broad at the base, tapering upward, and extending to the height of 150 or 200 feet. This pillar of flame revolved with an amazing velocity, while from its top proceeded a spire of black smoke, to a height beyond the reach of the eye, and whirling with the same velocity as the fiery column. During the whole period of its continuance, the column of flame moved slowly and majestically around the field. The noise of the whirlwind was louder than thunder, and its force so great, that trees six or eight inches in diameter, which had been cut, and were lying on the ground, were whirled aloft to the height of forty or fifty feet."

144. During the terrible conflagration of Moscow, in 1812, the air became so rarefied by the intense heat, that the wind rose to a frightful hurricane; the roar of the tempest being heard even above the rushing sound of the conflagration.

145. RESULTS OF CENTRIFUGAL ACTION. By the centrifugal action of the whirl, the air is driven *outward*, as in the case of hurricanes, and at the same time *spirally upwards*, on account of the pressure of the surrounding atmosphere: a great rarefaction, therefore, occurs at the centre. As long as the base of the whirlwind is *above* the ground, the warm air of the earth will stream *under* and *upwards*, into this partial void

---

Give the cases.
State the result of centrifugal action.

from every quarter; while, at the same time, the cold air will descend into it from the higher region of the atmosphere.

By this union, a powerful condensation of vapor occurs; causing the precipitation of rain and hail, and the development of electricity.

146. These, however, constitute no essential part of a whirlwind; for, if the currents of air that give rise to this phenomenon are very dry, the violence of the wind is the only remarkable circumstance. This was shown in the case of a small whirl, which involved two persons, who were going one cloudy day from Halle to Gie bichenstein. Suddenly they were separated by a gust of wind; one being driven against a wall, and the other thrown into a field; while the people who were near had not discerned the slightest disturbance in the atmosphere.

147. When the base of the whirlwind descends to the earth, it touches the surface, either partially or wholly, according as the axis is inclined or vertical. In the first case, the inward flowing currents will be *partially*, and in the second *entirely*, arrested by the centrifugal action of the storm.

The same results often occur when it covers a building. Hence, the atmosphere becomes exceedingly rarefied, both above, and around the edifice; and if it happens to be closed, and the tornado is violent, its walls will be burst *outward* by the sudden expansion of the air within, (C. 509.) Just as a sealed bottle of thin glass, under the exhausted receiver of an air-pump, is shivered by the elastic force of the enclosed air.

148. EFFECTS OF EXPANSION. In the tornado that happened at Natchez, in 1840, the houses exploded wherever the doors and windows were shut; the roofs shooting up into the air, and the walls, even of the strongest brick buildings, *bursting outward* with great

---

Are rain, hail, and electricity necessary to the production of a whirlwind?
Give the case.
Why are buildings burst outward by the action of tornadoes?
Give instances.

force; but no such destruction occurred when a free outlet was afforded to the air within. One gentleman as the storm approached, caused all the windows and doors of his house to be opened, and though its structure was frail it experienced no injury; not even a single pane of glass being broken.

149. On the 18th of June, 1839, a whirlwind (to which we have alluded) fell upon the village of Chatenay, near Paris. In the room of a house, over which it passed, several articles of needlework were lying upon a table: the next day some of them were found in a field, at a great distance from the house, together with a pillow-case taken from another room. They must have been carried up the chimney by the rush of air outwards, as every other means of exit was closed. Another singular illustration of the fact before us took place in the Shelbyville tornado. Soon after its occurrence, a lady missed a bonnet, which, the day before the storm, was lying enclosed in a bandbox in her chamber; some weeks afterwards, she accidentally observed a ribbon hanging from the chimney, which proved to be the string of her bonnet. The house had been closed during the storm, and the expansion of the air within the bandbox had forced off the lid—the lost article had been borne by the outward flowing current up the chimney, which afforded the only mode of egress, and there it had lodged.

CHAPTER IV.

WATER-SPOUTS.

150. A WATER-SPOUT *is a whirlwind over an expanse of water, as the sea or a lake, differing from a land-whirl in no other respect than that water is subjected to its action, instead of the bodies upon the surface of the earth.*

151. A water-spout usually presents the following

Define a water-spout.

successive appearances. At first it is seen as an inverted cone, either straight or slightly curved, extending downward from a *dark cloud* to which it seems to be attached. As the cone approaches the surface of the water, the latter becomes violently agitated, and, rising in spray or mist, is whirled round with a rapid motion. As the cone descends lower the spray rises higher and higher, until both unite, and a continuous column is formed extending from the water to the clouds.

The spout is now complete, and appears as an immense tube, possessing both a rotary and progressive motion; bending and swaying under the action of the wind as it advances on its course.

When the observer is near, a loud, hissing noise is heard, and the interior of the spout seems to be traversed by a rushing stream.

After continuing a short time the column is disunited, and the dark cloud gradually drawn up; for a while a thin, transparent tube remains below, but this at last is also broken, and the whole phenomenon then disappears. These successive changes are represented in figures 9, 10, 11, which are taken from sketches of water-spouts actually seen.

Fig. 9.

WATER-SPOUT FORMING.

What are its successive appearances?

70 AERIAL PHENOMENA.

Fig. 10.

WATER-SPOUT FORMED.

Fig. 11.

WATER-SPOUT ENDING.

152. FACTS. A water-spout occurred at Cleveland, Ohio, in September, 1835, which, from the description

---

Describe the one which was seen at Cleveland.

well illustrates the origin and characteristics of this phenomenon. "A heavy storm-cloud, driven by a *north-west gale*, was met by a strong *opposing current;* when an arm of the cloud appeared to drop down, and drag the waves up towards the sky. The whirling and dashing of the spray at the surface of the lake, and the column of water and mist extending, in a tall and tortuous line, to the cloud, were so well defined as to excite the admiration of all who observed them. At the expiration of about seven minutes, the north-wester triumphed, and swept the cloud to the south-east of the city."

153. The water-spout does not always pass through the various changes that have been detailed; sometimes the upper portion only is developed, depending from a mass of black clouds, like a huge, tapering trunk, without ever reaching the water; at other times, nothing is seen but the cloud of spray and mist that forms the base. On the voyage of the Exploring squadron from New Zealand to Tongataboo, a spout was beheld in the act of forming, at the distance of about half a mile. A circular motion was distinctly perceived, the water flying off in jets from the circumference of a circle, apparently fifty feet in diameter. A heavy, dark cloud hung over the spot, but no descending tube appeared, nor was there any progressive motion. In a short time the cloud dispersed, and the surface of the sea resumed its former state.

154. It is by no means uncommon for several water-spouts to appear at the same time. In May, 1820, Lieutenant Ogden beheld, on the edge of the Gulf stream, no less than *seven* in the course of half an hour: varying in their distance from the ship from two hundred yards to two miles.

155. DIMENSIONS. The *diameter* of the spout at its base ranges from a few feet to several hundred, and its

---

Does the water-spout always undergo these changes?
Under what forms is it sometimes seen?
How many have been seen at once?
What is the breadth and height of water-spouts?

altitude is supposed by some to be at times as great as a mile. In the account given by the Hon. Capt. Napier, of a spout which he beheld in 30° 47′ N. Lat., and 62° 40′ W. Lon., the diameter was judged to be 300 feet; and the height of the column to the point where it entered the hanging cloud, was computed, from observations taken by the quadrant, to be 1720 feet, or nearly one-third of a mile.

156. POPULAR ERROR. It is a common belief, *that water is drawn up by the action of the spout into the clouds;* but there is no proof, whatever, of a continuous column within the whirling pillar, and the fact, that the water, which sometimes falls from a spout upon the deck of a vessel at sea is always fresh, sufficiently refutes the idea. The torrents of rain, by which this phenomenon is often accompanied, can be fully accounted for by the rapid condensation of vapor that occurs, when the warm, humid air of the sea flows inward to the vortex of the whirl, and there combines with the cold air of the upper regions of the atmosphere, which descends to fill the partial void. From this union the electric phenomena of water-spouts arise, and the violent hail-showers that at times prevail; the mode, however, in which they originate, will be explained hereafter.

When a vessel is in the vicinity of water-spouts, cannon shots are usually fired for the purpose of destroying them; lest the vessel should be injured if a spout were to pass over it. It is not improbable that such an effect may be produced when the spout is either struck by the balls, or violently agitated by the concussion of the air arising from the discharges.

157. SAND PILLARS. Another form of the whirlwind is exhibited in the pillars of sand, which are not unfrequently seen in the deserts of Africa and Peru. Bruce, on his journey to Abyssinia, beheld *eleven* vast columns of sand of lofty height, moving over the plain at the

---

What popular error exists in regard to this phenomenon?
For what purpose are cannon discharged?
Where do sand pillars occur?

same time; some with a slow and majestic motion, and others with great velocity. Now, with their summits reaching to the clouds, they rapidly approached the terrified observers, and, the next moment, were borne away by the wind with incredible swiftness.

Their tops, at times, were seen separated from the main pillars, and the latter were often broken in two, as if struck by a cannon shot: the diameter of the largest was about ten feet.

While Mr. Adanson was crossing the river Gambia, a sand-whirl, twelve feet in breadth and two hundred and fifty in height, passed within forty yards of his boat.

158. The same phenomena are seen upon the Peruvian coast. "The sand," says Dr. Tschudi, "rises in columns from eighty to one hundred feet high, which whirl about in all directions, as if moved by magic. Sometimes they suddenly overshadow the traveler, who only escapes by rapid riding."

159. BENEFICIAL EFFECT OF WINDS. The utility of winds must be evident to all. By their aid vast oceans are crossed, and the products of distant climes wafted from shore to shore. Different nations are linked together by social and commercial ties, the blessings of civilization diffused, and the glad tidings from a better world borne to every land.

The growth and decay, both of animal and vegetable life, vitiates the atmosphere, and renders it unfit for respiration; but the winds prevent the deadly effects that would flow from this source, and the air becomes pure and salubrious, from its constant circulation.

Even the fierce tempest may be a messenger of mercy, by sweeping from the air the seeds of pestilence and contagion.

The advantage of winds in distributing moisture to the earth, will be seen in the following pages.

Describe those seen by Bruce and Adanson.
What does Dr. Tschudi relate?
What are the advantages arising from winds?

PART III.

AQUEOUS PHENOMENA.

CHAPTER I.

OF RAIN.

160. RAIN *is produced by the rapid union of two or more volumes of humid air, differing considerably in temperature;* the several portions *in union* being incapable of holding the same amount of moisture that each can *separately* retain.

This circumstance is the result of the law, that the capacity of the air for moisture decreases at a faster rate than the temperature.

161. This effect may be thus illustrated: 4000 cubic inches of air, at the temperature of 86° Fah., can contain no more than $31\frac{1}{2}$ grains of moisture, and an equal volume, at 32° Fah., only $7\frac{7}{8}$th grains. Now, if the two volumes are mingled together, their average temperature will be 59° Fahrenheit, and the weight of moisture they unitedly possess will be $39\frac{3}{8}$th grains. But, at this temperature, $31\frac{1}{2}$ grains is all the moisture that 8000 cubic inches of air can *possibly* retain; since the *first portion*, by its union with the *second*, diminished its capacity *one-half*, while that of the latter was only *doubled*. The excess, therefore, of $7\frac{7}{8}$ grains will be condensed, and descend in the form of water.

What does part third treat of?
What subject is discussed in chapter first?
How is rain produced?
Give the illustration.

162. Rain is the result of such combinations on an extensive scale, and the quantity that falls at any particular time or place, depends upon the difference in the several temperatures of the combining volumes, and the amount of moisture which each separately possesses.

163. Winds are the great natural agents by which such combinations are effected, and these occur most readily, when the currents of air are *shifting* and *variable*. Constant winds, blowing steadily from the same quarter and possessing an unchanging temperature, can produce no such admixture, and they are consequently attended with dry weather; except in the case where they strike the sloping sides of lofty mountains, carrying the warm air of the sea and the vales far up into the colder regions of the atmosphere.

164. RAIN-GAUGE. The quantity of rain that falls at any station during a given time, is ascertained by means of the rain-gauge, an instrument which is constructed in a variety of ways.

One of the best consists of a cylindrical, copper vessel, furnished with a float; the rain falling into the vessel raises the float, the stem of which is so graduated that an increase in depth, to the extent of one-hundredth of an inch, can easily be measured.

The greatest annual depth occurs at San Luis, Maranham, 2° 30' S. Lat.; and Vera Cruz ranks next in this respect. At the former place, 280 inches have been observed to fall in the course of a year, and at the latter 278 inches.

165. DISTRIBUTION OF RAIN IN LATITUDE. Since the capacity of the air for moisture increases with its temperature, we should naturally infer, that the higher

---

What circumstances influence the amount of rain which falls at any place?
What great natural agents effect these combinations?
What is said of variable and constant winds?
What is the rain-gauge?
Where does the greatest yearly depth of rain occur?
What is the law of distribution in respect to latitude?

the mean temperature of any region, the greater would be the amount of rain which descends upon it.

This is true as a general rule, for the annual depth of rain is found to *decrease* with the *increase* of latitude, as will be seen from the annexed list of seven localities, where the rain has been measured.

	North Latitude.	Annual depth of Rain in inches.
Grenada,	12°	126
Cape Francois,	19° 46′	120
Calcutta,	22° 35′	81
Rome,	41° 54′	39
London,	51° 31′	25
St. Petersburg,	59° 56′	16
Uleaborg,	65° 1′	13.5

106. EXCEPTIONS. Although this general relation to latitude exists, it is by no means to be supposed, that the same amount of rain descends yearly upon all regions lying within the same parallels; local causes will have their influence, and create, in many cases, extraordinary departures from the common rule. Thus, Bombay and Vera Cruz possess, nearly, the same position in latitude; but while at the former city, the annual depth of rain varies from 61 to 112 inches, that of the latter ranges from 120 to 278 inches.

This is owing to the following circumstances. Vera Cruz is backed by a chain of lofty mountains, rising beyond the limits of perpetual frost, and hither the hot and humid tropical air is constantly driven by the trade winds, as they sweep from the sea. Hence a great and sudden reduction of temperature occurs amid these icy regions, and the air, no longer capable of absorbing its vast stores of moisture, precipitates an immense quantity of rain.

167. At Bergen, in Norway, it has also been found, that in consequence of the moist south-west winds being checked in their course by the mountains, more than 88 inches of rain descend in a year: a quantity greater

---

Illustrate.
State the exceptions and the cause.

than that which falls at Calcutta during the same period.

168. DAYS OF RAIN. Though the annual amount of rain is greater in the low than in the high latitudes, the rainy days are not so numerous; as appears by the following table, which presents the average yearly number, within the latitudes indicated.

	N. latitude.	Mean annual number of rainy days.
From	12° to 43°	78
"	43° " 46°	103
"	46° " 50°	134
"	50° " 60°	161

169. From this circumstance we should readily conclude, that the *ordinary rains of the tropical climes must be more powerful than those of the temperate regions;* an inference which coincides with observation; for it is stated by Mr. Scott, that at Bombay, there once fell, during the first *twelve days* of the wet season, *thirty-two inches* of rain, a quantity equal to the *yearly average* of England.

170. This fact is also shown by the sudden rise of brooks and rivulets; a remarkable instance of which is related by Mr. Elphinstone, in the account of his mission to Cabul. It occurred between the Indus and Hydaspes, and is thus described. "On one occasion, the rear-guard, with some of the gentlemen of the mission were cut off from the rest by the sudden swelling of a brook, which had been a foot deep when they began to cross. It came down with surprising violence, carrying away some loaded camels that were crossing at the time, and rising about ten feet within a minute. Such was its force, that it ran in waves, like the sea, and rose against the bank in a ridge, like the surf on the coast of Coromandel."

171. DISTRIBUTION IN ALTITUDE. The great stores of atmospherical humidity reside in the inferior strata

---

What is the rule in respect to days of rain?
Where are rains most powerful?

of the air, and, for this reason, *less rain descends upon lofty table-lands and mountains, than upon regions situated lower down in the same latitude.*

Thus, in India, on the Malabar coast, twelve degrees from the equator, the annual depth of rain is 136 inches; while at Ootacamund, in the Nhilgerries, a region lying a short distance to the east, in the same latitude, but 8,500 feet above the ocean, the yearly amount of rain is only 63.88 inches. Likewise, at Sante Fe de Bogota, New Grenada, a city that enjoys an elevation of 8,800 feet, in the fourth degree of north latitude, the annual quantity of rain is nearly the same as that which falls in Germany, which is about *twenty-one inches.*

Even slight variations in altitude cause perceptible differences in the quantity of rain. At the Paris Observatory, a rain-gauge is placed in the court, and another upon the terrace, eighty-nine feet above. The mean annual depth of the rain which fell in the court for a space of ten years, was found to be 22.44 inches, and of that which descended upon the terrace during the same period, only 19.68 inches.

172. RAIN UPON COASTS. We have remarked, that the air above the ocean is always saturated, and that its humidity decreases as we penetrate from the sea-shore into the interior of a country. Conformably to this law, other things being equal, *more rain descends upon the coasts than upon the central regions of a country;* inasmuch as a less reduction of temperature will here produce a precipitation of moisture.

Besides, when the warm, humid air is borne inland by the winds from the sea, its course is marked by descending showers, and its inherent moisture decreases with its progress. Thus, on the west coast of England, 37 inches of rain fall in the course of a year; while in the interior, upon the eastern side, the annual depth is 25 inches. The maritime and inland regions of France

---

Give the rule in regard to distribution in altitude. Illustrate.
Compare the rain upon coasts and inland regions.
Why is there a difference? Give instances.

and Holland differ, in this respect, one inch. In this country, the yearly average fall of rain at Boston, for a period of 22 years, is 39.23 inches; at Hanover, New Hampshire, 38 inches; in New York State, 36 inches; and in Ohio, 36 inches. A diminution occurring as we advance into the interior, notwithstanding the influence of the great northern lakes, in the last two instances.

RAINS WITHIN THE TROPICS.

173. Upon the ocean, in the region of calms, where the gusts of wind are ever changing their direction, torrents of rain frequently descend. On the land, in all places where the trade wind blows constantly seaward, no rain falls, and the sky is always serene; but, wherever disturbances occur in this current and the monsoons prevail, *the rains are periodical, and the year is divided into two seasons, the wet and the dry.* These are so marked in their character, that whole months pass away without a cloud obscuring the sky, or mitigating the fierce heat of the sun: then the face of nature entirely changes, the heavens gather blackness, the rain comes down like a deluge, and the parched earth is refreshed, for many successive weeks, by copious showers.

174. RAINY SEASON. The rainy season commences, in all the countries within the tropics, at the *shifting of the monsoons;* and as this change is dependent upon the position of the sun, it begins earlier in those regions that lie near the equator, than in those more remote.

At Panama, 8° 48′ N. Lat., the rain falls early in the month of March; but it seldom appears at St. Blas, California, before the middle of June. In Africa, near the line, the rainy season begins in April, both upon the sea-coast and in the interior; but in the countries watered by the Senegal, it commences in June, and lasts till November.

How are rains distributed within the tropics?
How is the year divided where the monsoons prevail?
When does the rainy season occur?
In what regions early? In what late? Illustrate.

In India, the rains occur in May, at the southern extremity of the Malabar coast, but do not reach Delhi until nearly the end of June.

175. CAUSE. These stated rains originate in the *change of the periodical winds*, by which the union of vast volumes of air, differing in temperature, is rapidly effected. The subject cannot be better illustrated, than by recurring to the origin of the monsoons of India. (Art. 106.) Early in the month of June, the soil of the peninsula becomes intensely heated by the vertical rays of the sun, and powerful currents of rarefied air then ascend from the earth. To supply the deficiency thus created, the warm and humid atmosphere of the equatorial seas flows in, constituting the *south-west* monsoon; this wind now mingles with the cool, dry air, which the *north-east* monsoon, for the six previous months, has been constantly bringing to the peninsula from the polar and temperate, climes, and thus produces a combination favorable to the precipitation of rain, upon a most extensive scale.

176. PERIODICAL RAINS OF INDIA. On the Malabar coast, the south-west monsoon is ushered in by terrific storms of thunder and lightning, the water pours down in torrents, and, when the thunder has ceased, nothing is heard for several days but the rush of the descending rain, and the roar of the swelling streams. In a few days, the storm ceases, and the earth, which before was withered by the glowing atmosphere, is now, as if by magic, suddenly clothed with the richest verdure; the air above floats pure and balmy, and bright tropical clouds sail tranquilly through the sky.

After this, the rains fall at intervals for the space of a month, when they again return with great violence. In July, they attain their height, and from that time gradually subside until the end of September, when the season closes, as it began, in thunders and tempests.

177. The following table, the result of the observa-

How do these rains originate?
Describe those of India.

tions of twelve years, shows the mean monthly average for the rainy season, at Bombay; and serves to elucidate the preceding remarks.

	Inches.
June,	24
July,	23.95
August,	18.87
Sept.,	14.06
Oct.,	1.06

178. The south-west monsoon does not, however, bring rain to the whole of India. Parallel to the *west-ern* coast runs a chain of high mountains, termed the Ghauts: here the monsoon is arrested in its course, and most of the moisture with which it is charged, is precipitated, ere it arrives at the central table-land of Mysore. On the *eastern*, or Coromandel coast, its influence is not felt, and the seasons are here *reversed*. From March till June, the winds are hot and moist, blowing mostly from the south, over the Bay of Bengal; from June to October the heat is very great, but about the middle of the latter month, the cool, northeast monsoon commences, bringing the periodical rains, which terminate by the middle of December; the monsoon continuing to blow until the beginning of March.

179. PERIODICAL RAINS OF CONGO. We trace the rainy and dry seasons of Congo, in the southern hemisphere, to the same cause. In general, from about March to September, no rain descends, but gales from the *south* and *south-east* temper the burning atmosphere. In October, hot and humid winds blow from the *north-west* over the Gulf of Guinea, and the country is then flooded by frequent rains, which continue to increase until January. Slight showers then fall at intervals until March, when the rains recommence and continue for a short time.

Illustrate from the table.
What is the influence of the Ghauts upon the south-west monsoon?
What is said of the seasons on the eastern coast?
What wind brings the rains to this region?
Describe the periodical rains of Congo.

RAINS IN THE HIGHER LATITUDES.

180. *Beyond the tropics, the rains no longer occur at stated periods, but are distributed throughout the seasons without regard to any law.*

Thus, in the west of England, the amount of rain in winter is *eight times greater* than in summer; but in Germany, it is *one-half* of what falls in summer, and at St. Petersburg a little more than *one-third*. In Italy the greatest quantity descends in autumn. There is the same irregularity in the number of rainy days; for in the west of England, there are more rainy days in winter than in summer; but in Siberia, it rains four times as often in summer as in winter.

181. RAINY WINDS. The rains in the higher latitudes, as well as within the tropics, depend upon the changes of the wind; though one wind may be more productive of rain than another, and, in different regions, the rainy winds do not always blow from the same direction.

In Europe, north of the Alps, the north-east wind is dry and cold, since it sweeps over the land from the higher latitudes; but the south-west wind brings the rain, for, coming over the Atlantic from southerly climes, it is warm and humid, and its capacity for moisture is constantly decreasing.

Out of *one hundred* showers that were noted at Berlin, scarcely any occurred when the north-east wind prevailed; while nearly half were brought by the winds from the south-west and west. Moreover, it rained only *once* for every *nine times* that the *easterly winds* blew, but *thrice* for the same *number* of times in which the *south-westerly* breezes predominated.

182. The reverse of this occurs on the eastern coast of the United States, for here the north-east winds give rise to the long storms of the fall and spring. At these seasons, as appears from the observations of Dr. Hale,

---

Where are the rains irregular? Give cases.
What is the rainy wind of Northern Europe?
Why is it rainy? Give instances.
Whence comes the rainy wind, on the eastern coast of the United States?

of Boston, continued through a period of twenty-two years, the winds are colder than the atmosphere of the land, and as they come from the sea charged with moisture, the cause of the rain is readily discerned.

REGIONS WITHOUT RAIN.

183. EGYPT. In Egypt it scarcely ever rains. At Cairo, there is an average of four or five showers a year; but, as we recede from the coast, it becomes more rare, until in Upper Egypt, under the cloudless sky of Thebes, a man's life may pass away without his ever beholding a single rain.

184. The cause of this scarcity of rain is to be sought in the *peculiar conformation of the surface of this country.* It is a narrow valley, bounded by two mountain ridges on the *east* and *west;* the first prevents the moisture exhaled from the Red sea from reaching the valley, and, as the African deserts extend beyond the western range, no source of rain exists in this quarter.

185. The northerly winds, which blow from May till October, bearing off the vapors of the Mediterranean, pass over the whole length of the valley of the Nile, without meeting any obstruction; and it is only when they are driven up the high range of the Abyssinian mountains, that they become sufficiently cooled to precipitate rain. Here it descends most copiously during the summer months, swelling the tributaries of the Nile, and producing its annual inundation.

186. Much of the humidity brought by these constant winds, can be retained by the atmosphere of Egypt, without being precipitated; since it is far below the point of saturation, in consequence of the prevalence of hot, dry winds from the desert, (Art. 114,) and the extreme aridity of the soil.

So free from moisture is the ground, that myriads of human bodies have rested for centuries within its bosom

---

Why is it rainy?
What is said of Egypt?
Why is it that rain rarely falls in this country?
What is said of the dryness of the soil?

without suffering the least decay; and in a collection of antiquities, now in the British Museum, there is an ancient model of an Egyptian house, the store-rooms of which, when first discovered, were full of grain that had remained uninjured for ages.

187. PERU. Along the coast of Peru is stretched a plain of sand, five hundred and forty leagues in length, and varying from three to twenty in breadth, upon which no rain descends. So rare is the occurrence of a real shower at Lima, that it is a source of terror; and when such an event happens, religious processions parade the streets, imploring the protection of heaven for their endangered city.

The want of rain in this region is thus explained. Parallel to the coast of Peru, and at a short distance from the sea, extends the lofty range of the Andes, the peaks of which rise far above the limit of perpetual frost. The constant east wind, sweeping from the Atlantic to the Pacific, across the extreme breadth of South America, gradually ascends the slope of the Andes; but by the time it has passed their summits most of the vapors with which it is charged, are precipitated, and it comes to the shores of Peru comparatively destitute of moisture.

188. Moreover, as a sandy soil is naturally dry, scarcely any evaporation occurs, and the hot air of the plains possesses but little humidity. For these reasons, the difference in the temperature of two or more combining volumes of air is rarely sufficient to produce rain.

189. A similar destitution of rain exists on the northwest coast of Africa, where the desert of Zahara reaches the Atlantic. In this region, intervals of *six or seven* years occur between the showers.

190. CONSTANT RAINS. In Guiana, it rains for a

What is said of Peru?
Is a copious shower regarded as a blessing at Lima?
Explain the cause of this scarcity of rain.
What other region is destitute of rain?

great portion of the year; nor is this surprising, when we reflect that this country is a low and marshy region, overspread with immense forests; situated but a few degrees north of the equator, and subjected to the influence of the north-easterly trade.

The fierce heat of the sun fills the atmosphere with vapor, which returns to the earth again in incessant showers, as the cool air from the ocean flows in from the higher latitudes.

In the interior, amid the primeval forests, the sun and stars are seldom visible, and the rains not unfrequently continue for *five or six months*, with scarcely any intermission.

191. According to Darwin, rain thus prevails at the Straits of Magellan. "At Port Famine," says the writer, "we have rounded mountains, concealed by impervious forests, which are drenched with rain brought by an endless succession of gales: rock, ice, snow, wind and water, all warring with each other, here reign in absolute sovereignty." It is a proverbial saying, in the Isle of Chiloe, 43° S. Lat., that it there *rains six days* of the week, and is *cloudy* on the *seventh*.

192. EXCESSIVE SHOWERS. The quantity of rain that falls during a single shower is sometimes amazing. At Cayenne, Admiral Roussin found, on one occasion, that ten inches and three quarters fell in the course of ten hours. There fell at Genoa, Oct. 25th, 1822, thirty inches in twenty-four hours; and at Geneva, on the 20th of May, 1827, six inches in three hours. In the famous Catskill storm of July 26th, 1819, a tub, very nearly as large at the bottom as at the top, was filled to the depth of fifteen inches and a half in *four hours*.

193. RAIN WITHOUT CLOUDS. Singular as it may appear, there are yet many well-attested instances of showers occurring when the sky was *clear*. This phenomenon was several times observed by Humboldt; and,

---

What is said of the rains of Guiana? What of those at Port Famine?
Give instances of excessive showers.
Does rain ever fall from a cloudless sky?

according to Kaemtz, it happens in Germany twice or thrice in a year. On the 9th of August, 1837, a shower fell at Geneva, when the sky was cloudless, that lasted two or three minutes; and at Constantinople, rain was seen to fall by M. de Neveu, for the space of ten minutes, when the heavens were perfectly serene. According to Le Gentil, this occurrence is by no means uncommon in the island of Mauritius, during the prevalence of the south-east winds; slight showers falling in the evening, when the stars are shining brilliantly.

194. CAUSE. The following explanation has been given of this phenomenon. When rain is produced by the intermixture of different volumes of air, the precipitated moisture usually assumes, at first, the form of small globules of vapor; an assemblage of which in the higher regions of the atmosphere constitutes clouds. As the process of condensation advances, more moisture is precipitated, and the globules uniting in rain-drops, descend to the earth. Now it is supposed, that, at times, the humidity of the atmosphere is condensed *at once* into rain, without passing through the intermediate state of cloud; and under these circumstances a shower might fall from a cloudless sky.

CHAPTER II.

OF FOGS.

195. *Fogs, or mists, are visible vapors, that float in the atmosphere, near the surface of the earth.*

They originate in the same causes as rain; viz., the union of a cool body of air with one that is warm and humid; when the precipitation of moisture is *slight*, *fogs* are produced; when it is *copious*, *rains* are the result.

Give cases. How is this circumstance explained?
What is the subject of chapter second? Define fogs.
In what do they originate?

196. Constitution. When a mist is closely examined, it is found to consist of *minute globules*, and the investigations of Saussure, and Kratzenstein, lead us to suppose, that they are *hollow ;* for the latter philosopher discovered upon them rings of prismatic colors, like those seen upon soap bubbles; (C. 79,) and these could not exist if the globule was a drop of water, with no air or gas within. The size of these globules is greatest when the atmosphere is very humid, and least when it is dry.

DISTRIBUTION IN LATITUDE.

197. Tropical Regions. Fogs are not generally common in the equatorial clime, its high mean temperature being favorable to the dissolution of vapor. They are however, by no means, unfrequent at certain seasons, and in particular localities.

Thus, in India, just before the commencement and at the close of the rainy season, when the air contains an excess of moisture, but not enough to produce rain, clouds of mist so dense and thick obscure the atmosphere, that they are not dissipated until late in the morning.

During the month of December, the towering summits of the Abyssinian mountains are also shrouded in impenetrable fogs. Peru is remarkable for its misty atmosphere, of which we shall soon speak more particularly.

198. Temperate Regions. In the temperate climes, mists *frequently* occur ; but are of comparatively *small extent*.

199. Polar Regions. In the polar regions they spread *far* and *wide*, over sea and land, and prevail both in winter and summer.

At the beginning of winter, the whole surface of the northern ocean steams with vapor, denominated *frost smoke;* but as the season advances, and the cold in-

---

What does a mist consist of?
Where do fogs prevail least ? When do they appear in India ?
Where do they occur frequently ? Where most ?

creases, it disappears. Towards the end of June, when the summer commences, the fogs are again seen, mantling the land and sea with their heavy folds. By the middle of summer, these also disappear, to return again at the approach of winter.

So dense are these mists, that they render the navigation of the polar seas extremely dangerous, and the narratives of the hardy explorers of these inhospitable climes are full of the perils arising from this source. Simpson, who penetrated by land to the Arctic ocean, in 1837, speaks of the dense fog that often involved his party in imminent danger, while coasting along these ice-bound shores.

200. CAUSE. The phenomena of the polar fogs are explained in the following manner. During the short Arctic summer, the earth rises in temperature with much greater rapidity than the sea: the thermometer sometimes standing, according to Simpson, at 71° Fah. in the *shade*, while ice of immense thickness lines the shore. Flowers also bloom at the surface of the ground, when the soil is firmly frozen *four inches below*. The air, incumbent upon the land and water, partakes of their respective temperatures; and on account of the ceaseless agitations of the atmosphere, a union of the *warm* air of the *ground* with the *cool* air of the *ocean* will necessarily occur, giving rise to the *summer* fogs. But, as the winter approaches, the *land* becomes *colder* than the *sea*; since the heat acquired during the season of summer is lost far more slowly by the latter than by the former; and then, upon the warm surface of the ocean, will float the *frost smoke*, as the cool air flows down upon it from the adjacent shores.

LOCAL DISTRIBUTION.

201. *Fogs are found along the course of rivers, upon the sides of mountains, and over shoals and capes.* It is not difficult to detect the cause of their appearance in these situations.

Describe the polar fogs. Explain the cause of their formation. In what localities are mists found?

LOCAL DISTRIBUTION.

202. RIVERS. The banks of a river, during the night, lose more heat by radiation than the stream itself, and to the air, which rests upon each, a similar difference in temperature is imparted. By the fluctuations of the atmosphere, an intermixture is readily effected; and the superfluous moisture is seen, in the morning, floating in fog over either bank, and tracing in a wreath of mist the devious windings of the stream. Fogs usually occur over rivers in the *early* part of the day; for the reason, that soon after the sun rises the equality of temperature is restored, and the vapor is then rapidly dissipated.

203. When Sir Humphrey Davy descended the Danube in 1818, he observed that mist was regularly formed, when the temperature of the air on *shore* was from *three* to *six degrees lower* than that of the *stream;* and, at the junction of the Inn and Ilz with the Danube, at six o'clock on a morning in June, he found the distribution of temperature, and the state of the mist, to be as follows.

Temperature of the air over the land.	Temperature of the rivers.	State of the atmosphere over the rivers.
54° Fah.	Danube, 62° Fah. Inn, 56° Ilz, 55°	Thick fog all over. Dense mist all over. Light mist.

204. It is not essential to the production of fogs, that the air of the stream should be *warmer* than that of the land; it may be *colder*, and then fogs appear, if the difference of temperature is sufficiently great. This is the case on the Mississippi. During the spring and fall, mists form over the river in the *day time*, when the temperature of the water is several degrees below that of the air above, and the air above cooler than the atmosphere upon the banks. These diurnal fogs, though often extremely dense, are chiefly confined to the river, and seldom extend beyond its banks.

Why do they occur along the course of rivers?
State Sir H. Davy's observations.
Under what other circumstances can mists occur? Give instances

205. On the 31st of Dec. 1847, as the writer was standing upon a bridge, which crosses one of the tributaries of the Connecticut, he was unable to perceive a mill, 140 yards distant, in consequence of the dense fog which covered the river. Upon examination, the temperature of the water was found to be 32° Fah., and that of the air, close by the stream, 46° Fah.: a difference, here existing, of *fourteen degrees.*

206. MOUNTAINS. Fogs appear upon mountains, because the warmth of the atmosphere diminishes as we ascend, (Art. 51,) and the cool and shady forests, that clothe their sides, contribute still further to lower the temperature. Hence, when the warm air of the vales is gradually driven up by the wind into these regions, its capacity for moisture is continually reduced, until at length a precipitation occurs, and clouds of mist involve both cliff and forest.

207. At the Mountain-House, on the Catskill range, the temperature in summer is *ten degrees lower* than in the valley of the Hudson: and often when a breeze sets towards the mountain, a spectator upon the summit beholds, at first, a wreath of mist extending along the base; soon the lower belt of forest is concealed from view, and the fog continuing to ascend, thickening and spreading on every side, the landscape ere long is completely veiled, and a chilling wind sweeps past, loaded with moisture. A fact related by Sir John Herschel, affords a striking illustration of the influence of trees in condensing moisture. During the residence of this gentleman at the Cape of Good Hope, he observed, that on the side of the Table-Mountain from which the wind blew, the clouds were spread out and descended very low, and often without any rain falling; while on the opposite side they poured over the face of the mountain in dense masses of vapor. Sir John discovered, when walking beneath tall fir trees, while these clouds were closely overhead, that he was subjected to a copious *shower*, but on coming from beneath the trees, the rain ceased. On

---

Why do fogs appear upon mountains? Illustrate.
What fact is related by Sir John Herschel?

investigating the cause, he found that the clouds were condensed into rain on the cool tops of the trees.

208. CAPES. The reason for the existence of fogs over capes and headlands has already been given, in accounting for the prevalence of mist in the polar climes.

The soil of these places becomes *warmer* in *summer* than the ocean that washes their shores; but in the *winter, colder;* and the difference in temperature is usually sufficient to produce a constant succession of mists.

209. SHOALS. A similar state of the atmosphere occurs over shoals, inasmuch as their waters are colder than those of the main ocean. Thus, Humboldt found near Corunna, that while the temperature of the water on the shoals was 54° Fah., that of the deep sea was as high as 59° Fah.

Under these circumstances, an intermixture of the adjacent volumes of air, resting upon the waters thus differing in temperature, will naturally occasion fogs.

210. NEWFOUNDLAND. Mists of great extent shroud the sea on the banks of Newfoundland, and particularly near the current of the Gulf Stream. The difference in the warmth of the waters of the *stream*, the *ocean*, and the *banks*, fully explains this phenomenon. This current, flowing from the equatorial regions, possesses a temperature $5\frac{1}{2}°$ Fah. above that of the adjacent ocean, and the waters of the latter are from 16° to 18° warmer than those of the banks. The difference, in temperature, between the waters of the stream and banks, has even risen as high as *thirty degrees*.

211. ENGLAND. England, surrounded by a warm sea, is subject to thick fogs, that prevail extensively in the winter. In London they are often so extremely dense, that it is necessary to light the gas in the streets and houses in the middle of the day. On the 24th of

---

Why over capes? Why over shoals?
How are the fogs of Newfoundland explained?
Describe those of England.

February, 1832, people in the streets were unable, at mid-day, to see distinctly on account of the fog; and in the evening, the city having been illuminated, as this day was the birth-day of the queen, boys went about with torches, saying, "that they were looking for the illumination." Similar fogs have been observed at Paris, and Amsterdam. The smoke, arising from the extensive combustion in such large cities, is regarded, by some, as contributing to the density of these extraordinary fogs.

212. GARUAS. We have seen, (Art. 187,) that along the coast of Peru, the atmosphere scarcely ever possesses sufficient moisture to produce rain; it contains, however, enough to create widely extended and continued fogs.

The wintry season, in this country, lasts from April to October, and, throughout the whole of this period, a veil of mist shrouds sea and shore. At the beginning and end of this season, it rises between nine and ten in the morning, and disappears about three in the afternoon, at the hottest portion of the day. But, during the months of August and September, the vapor is extremely dense, and rests *for weeks immovably* upon the earth. In October and November, the misty canopy begins to rise, and gradually growing thinner, at length yields to the piercing rays of the sun, and is entirely dissipated.

213. These fogs, termed by the natives, *Garuas*, are said to be at times so heavy, that the moisture falls to the earth in large drops, which are formed by the union of small globules of mist. There is, however, this distinction, between them and rain-drops; that the latter descend from the more elevated regions of the atmosphere, while the *garuas* do not extend higher than *twelve hundred feet;* their average altitude varying from seven to eight hundred.

214. Passing eastward from the Puna table-lands of Peru, across the lofty ridges of the Andes, the traveler,

---

Describe the Garuas.
Describe the state of the atmosphere east of the Puna regions.

after descending a few hundred feet, arrives at a region totally different from that which he has just left. He no longer breathes a pure and refreshing atmosphere; for the air is loaded with vapors, and the wooded ranges, called the Ceja de la Montaña, or *mist of the mountain*, are clouded with fogs throughout the year. In the dry season, these are dissolved, during the day, by the powerful influence of the sun; but in the winter they condense upon the hills, and descend in ceaseless torrents of rain.

215. Proceeding in the same direction, from the Ceja de la Montaña, the magnificent slope of the Andes soon opens upon the sight; not reposing beneath a clear and azure sky, but overshadowed by a thick veil of mist, impenetrable to the rays of the early sun, and yielding only to his noontide beams.

216. The explanation of these phenomena is to be found, in the constant advance of the humid trade wind, from the eastern shores of South America to the towering summits of the Andes. Rising continually in its onward progress into higher and colder regions, its capacity for moisture is ever diminishing, and the atmosphere is always near the point of saturation. Its inland course will thus be marked by abundant rains, and when these abate, fogs and mists succeed in their turn.

By the time this great aërial current has arrived at the more elevated ridges, most of its humidity has been discharged; during the dry season enough only remains to produce extensive mists; and when, at length, it has reached Peru, it possesses scarcely any moisture. (Art. 187.)

217. A powerful auxiliary cause exists, in the rich and luxuriant vegetation, that springs up every where throughout this boundless region. The light of a tropical day, in its meridian splendor, can scarcely pierce the massive foliage of those mighty forests, which stretch away for leagues from the base of the lower Andes; while the lighter forms of vegetation, spreading in wild

Explain the cause.
What is the effect of the vegetation in this particular?

exuberance over the higher belts, effectually shield the earth from the fierce rays of the sun, and check the progress of evaporation. The soil, thus shaded, is always moist, and the air warm and humid; and from the causes already stated, such results are here produced as we should readily infer—*excessive rains in the lower forests* (190), and *clouds of mist* upon the more *elevated ranges*.

CHAPTER III.

OF CLOUDS.

218. *The name of clouds is given to those collections of vapor, that float at a lofty altitude above the earth.*

219. Though differing from fogs in *situation*, they originate in precisely the same causes; being formed, in the higher regions of the atmosphere, by the union of warm and cold air, when the combining volumes are over saturated. The excess of humidity, when slight, then appearing in the atmosphere in the form of *clouds*.

220. During the daily process of evaporation, warm, humid currents of air are continually ascending from the earth; the higher they ascend, the colder is the atmosphere into which they enter; and, as they continue to rise, a point at length will be attained, where, in union with the colder air, their original humidity can no longer be retained; a cloud will then appear, which increases in bulk with the upward progress of the current into colder regions.

If the cloud however, in its ascent, either meets with a warmer stratum of air, or descends towards the earth into a region of a higher temperature, a portion of the

---

What is the subject of chapter third?
Define clouds.
How do they originate?
In what manner do ascending currents produce clouds?

minute globules of water which compose it, perhaps all, will be re-dissolved, and the cloud will either contract in size, or completely vanish, according to the increase of heat to which it is subjected.

221. The entire atmosphere, to the altitude of many thousand feet, is constantly traversed by numerous horizontal currents of air, flowing in different *directions*, and at different *heights*. Combinations of vast volumes of air, varying in *temperature*, must therefore at times inevitably occur, as well in the higher as in the lower regions of the atmosphere; and when the excess of moisture resulting from this union is but *small*, clouds, with their ever-changing forms, obscure the serenity of the sky.

222. When Clayton, on the 31st of July, 1837, ascended in a balloon from Louisville, Ky., the direction of his course was altered no less than *five times*, in the space of *fourteen hours*. Once, when at a very great height, he beheld, a mile below him, a snow-white cloud of a mountain shape drifting in an *opposite* direction to that in which he was traveling.

223. The upward impulse given to the warm atmosphere near the earth, when driven by the wind against the sloping sides of mountains, is also a fruitful source of clouds. (230.)

224. STRATA OF CLOUDS. When an extended range of clouds settles down towards the earth, its under surface often copies the outline of the landscape immediately beneath, assuming a horizontal direction. This is owing to the high temperature of the air below the cloud, in consequence of which the latter would cease to be visible, were it to descend lower; for the globules of vapor would then be dissolved by the warm atmosphere.

225. Above the first range of clouds, the temperature is often much higher than in the region of vapors be-

---

What is said as to the existence of horizontal currents, and their effects?
Relate the illustration.
What is the influence of mountains?
What is said as regards the figure of the under surface of clouds?
How is the existence of successive ranges of clouds explained?

neath. Here the air will be clear, and a tract of considerable thickness frequently intervenes before we arrive at a second range of clouds; to this may succeed another body of pure air, and still higher a third range of clouds, and so on, alternately.

226. The following account, given by Jolliffe, of his aërial voyage, which took place in England, in 1826, is instructive in this connection.

"Our progress, during the first quarter of a mile, was so gradual, as to be nearly imperceptible; but on discharging a portion of the ballast, the balloon ascended with a rapidity, which, in a few minutes, buried us in the vapors of a dense mass of clouds. The temperature was here cold and raw; such as I have felt on a mountain-top, when enveloped in fog. We loitered here for some time; but at length rose with uncontrollable velocity, and burst, almost suddenly, out of this dark barrier into realms of light and glory. The stratum of clouds from which we had emerged, seemed depressed to a vast distance below us, involved in radiant folds, which completely shut out all view of the earth."

227. THICKNESS. The thickness of clouds is sometimes immense. On the 29th of Sept. 1826, Peytier and Hossard, two French engineers, were upon the Pyrenees, and so stationed, that they beheld, at the same time, the *upper* and *lower* surfaces of the *same cloud*. As the altitude of each station was known, the thickness of the cloud was readily determined, and found to be 1,476 feet. On the succeeding day, the thickness of the clouds was 2,788 feet; or more than *half a mile*.

228. HEIGHT. The height of clouds has been variously estimated. According to observations given by Dalton, *two-fifths* of all the clouds observed in England for the space of five years, were *more than* 3,150 *feet* above the earth. By noting when the upper and lower surfaces of the clouds touched the peaks of the Pyre-

---

Relate the account given by Jolliffe.
What is said respecting the thickness of clouds?
What of their height?

nees, which had previously been measured, Peytier and Hossard obtained no less than *forty-eight* altitudes. It was thus found, that the *lower surfaces* here varied in height from 1,476 feet to 6,200, and the *upper* from 2,952 feet to 9,810.

229. The computations of many distinguished observers have been collected by Kaemtz; and from these it appears, that clouds range, in height, from 1,300 feet to 21,320.

The extreme elevation here given is, however, not sufficiently great; for clouds are sometimes seen floating above the summit of Chimborazo, which rises 21,480 feet above the sea-level; and when Gay Lussac, in the month of September, 1804, ascended in a balloon to the altitude of 23,000 feet, he beheld clouds still soaring above him, apparently at a great height.

30. CLOUDS ON MOUNTAINS. When a mountain range is viewed from a distance, the various peaks are frequently seen capped with a cloud; while the atmosphere between them is perfectly clear. This appearance sometimes continues for hours, and even entire days; and was often noticed amid the Alps by the celebrated Saussure. It is caused by the wind impelling up the sides of the peaks the warm, humid air of the vales, which, in its ascent, gradually sinks in capacity, until it is over-saturated, when the excess of moisture becomes visible, and appears as a cloud.

231. This phenomenon is illustrated by figure 12. Let A B C represent the outline of a mountain peak, up the sides of which a warm current flows, in the direction of the arrows. Above the line D E, the temperature is below the dew-point of the current, and its humidity is condensed into a cloud at B. As the wind sweeps over the summit, the cloud B is carried below the line D E, on the opposite side, and re-dissolved in the warm atmosphere beneath; but its place, meanwhile, is occupied by a fresh cloud, caused by the ascent

---

What is the appearance sometimes presented by distant mountains?
How is this accounted for?

AQUEOUS PHENOMENA.

Fig. 12.

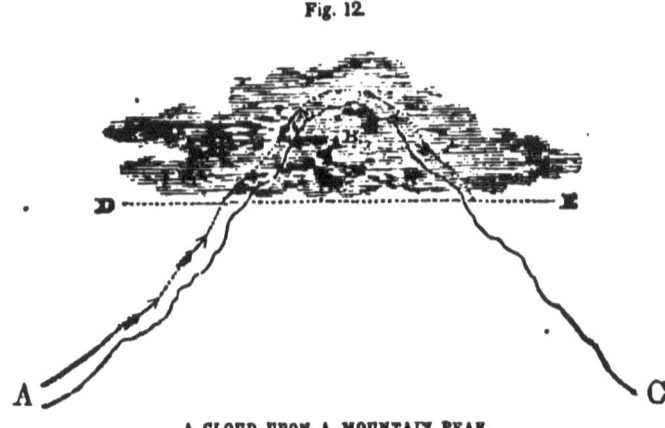

A CLOUD UPON A MOUNTAIN PEAK.

of the warm air on the side A B. It thus occurs, that though the cloud upon the mountain is *stationary* for hours together, yet the *particles* which compose it are continually *changing*.

232. The appearances just described are finely displayed upon the St. Gothard, a mountain in Switzerland, about 6000 feet high. Dark, heavy clouds that have formed on one side of the mountain, are frequently seen, passing rapidly over its summit, and descending in dense masses into the vale of Tremola, on the opposite side; but, instead of filling the plains beneath with thick vapor, the clouds are dissolved by the warm air into which they are precipitated.

233. A singular instance of the alternate appearance and disappearance of a cloud occurred, not long since, upon the coast of England. A cloud was seen, borne along by the wind, apparently passing from one side of an arm of the sea to the other, but not extending across the water. It was visible over the *land*, on each shore, but the *sky* above the water was perfectly *serene*. This phenomenon may be thus explained. Over the land, in the region of the cloud, the air was *below* the *dew-point;* but over the water, the sea being warmer than the land,

---

Explain from the figure.
Give the illustrations.

the temperature of the air was higher, and *above* the *dew-point.* When, therefore, the wind carried the cloud over the sea it *vanished*, its moisture being re-dissolved by the atmosphere; but when the body of air in which the cloud had previously existed, arrived at the opposite shore, a second precipitation of moisture took place, and the cloud *reappeared.*

CLASSIFICATION.

234. Clouds have been divided into *seven kinds; three original;* viz. the *cirrus*, the *cumulus*, the *stratus ;* and *four formed by combination*, viz. the *cirro-cumulus*, the *cirro-stratus*, the *cumulo-stratus*, and the *nimbus.*

235. CIRRUS OR CURL CLOUD. This cloud is so called, from the Latin word *cirrus* or *curl*, because it usually resembles a distended lock of hair. It is distinguished from the other kinds by its fibrous structure, the lightness of its appearance, and the variety of figures it is capable of assuming. After a period of fine weather, slender filaments of the cirrus are frequently seen, stretching like white lines across the azure sky. Sometimes these threads of clouds are arranged in parallel bands, which in the northern hemisphere, (wherever observations have been taken,) are either directed from south to north, or from south-west to north-east; at other times they separate, resembling the tail of a horse; a form which is known in Germany by the name of *wind-trees.* These filaments are also not unfrequently seen crossing each other, and investing the sky with a delicate net-work of gauze-like vapor. One of the most beautiful forms of the cirrus occurs, when the fibres curl from each side of a band of light cloud, and the whole appears like the feathered grain of a rich piece of mahogany, (figure 13, *a.*)

236. The white color of the cirrus renders it difficult, in all cases, to detect its peculiar structure; for the eye

---

Into how many classes are clouds divided?
What are they?
Describe the cirrus.

is dazzled by its excessive light. The cloud may, however, be viewed at leisure, by reflection from a blackened mirror, which diminishes the brightness.

237. The *cirrus* soars the *highest* of all clouds. Its altitude, at Halle, in Germany, has frequently been estimated, by Kaemtz to be not less than 21,300 feet; and, from the observations of ten years, this distinguished meteorologist has been led to believe, that it is entirely composed of *snow-flakes*. Indeed, the temperature of the elevated regions in which it floats, must be often far below the freezing point.

238. CUMULUS. This kind of cloud acquires its name from the Latin word *cumulus* or *heap;* the vapor seeming to be *piled* or *heaped* together. It is usually seen in the form of a hemisphere, resting upon a horizontal base; but at times detached masses gather into one vast cloud upon the horizon; their radiant summits gleaming like the snowy peaks of distant mountains, (figure 13, *b*.)

239. The *cumulus* is the *cloud of day*, and is produced by the ascending currents of warm air, caused by the solar heat. During the fine days of summer, its peculiar figure is most perfect, and its formation and decline occur in the following manner. Although the sun may have arisen in a cloudless sky, a few solitary specks of vapor may be seen towards eight or nine o'clock; these, as the day advances, enlarge from within, become thicker, and accumulate in rounded masses, which continue to increase in number and size, till the hottest part of the day. After this time they gradually lessen, and often entirely vanish, leaving the sky at sunset again perfectly serene.

240. The cumulus floats low in the morning; but its

How may its peculiar structure be best discerned?
How far above the general surface of the earth does the cirrus rise?
Of what does it consist according to Kaemtz?
Describe the cumulus.
How does it originate?
Describe the mode of its formation and the changes it undergoes?

altitude increases with that of the ascending currents, which attain their highest elevation soon after mid-day; towards evening the currents subside, and the cloud descends. This circumstance has often been remarked by meteorologists, when stationed on elevated mountains. In the *morning*, the cumulus has been seen *beneath* them; it enveloped them towards *noon;* then soared *above* them for several hours, and *descended* to the vale at the *close of day*.

Fig. 13.

CIRRUS (*a*), CUMULUS (*b*), AND STRATUS (*c*).

What is said of its height in the morning, at mid-day, and in the evening?

Account for the facts stated in ¶ 240.

241. It is not difficult to account for the facts just detailed. The cumulus begins to be formed, when the warm currents, in their upward progress, arrive at a temperature so low that they become over-saturated with moisture; and the excess is then condensed into a cloud.

The higher the currents rise, the colder is the atmosphere, generally speaking, and the cloud must necessarily enlarge; but when in the afternoon the strength of the currents abates, the clouds which are buoyed up by their force, sink down into warmer regions of the atmosphere, and are either partially or completely dissolved.

242. The *rounded figure* of the cumulus is attributed by Saussure to the mode of its formation; for when one fluid flows through another at rest, the outline of the figure assumed by the first will be composed of *curved lines*. This may be seen, by suffering a drop of milk, or ink, to fall into a glass of water; but the shape of a cloud of steam, as it issues from the boiler of a locomotive, presents a far better illustration.

243. STRATUS. This cloud derives its name from the Latin word *stratus*, or *covering*; it *forms* about *sunset*, *increases* in density during the *night*, and *disappears* at *sunrise*. It is caused by the vapors which have been exhaled during the heat of the day, but return again to the earth towards the evening, when the temperature has declined, and are then condensed into a sheet of clouds, which stretch along and rest upon the horizon (figure 13., *c*). This class likewise includes those light and spreading mists, which gather in meadows and vales in the evening of a warm summer's day, floating like a veil over the surface of the ground, and extending but a short distance above it.

244. CIRRO-STRATUS. This cloud is so called, be

What causes the rounded figure of the cumulus?
Describe the stratus.
When does it form, increase and vanish?
How does it originate?

cause it partakes of the characteristics of the *cirrus* and *stratus;* originating usually in the cirrus. It is remarkable for its great length, in proportion to its thickness; but though preserving in the main this peculiarity, it assumes many varieties of form.

245. At one time it consists of a number of parallel *bars of vapor,* in close proximity, blended together at the middle, but separated at the edges (figure 14., *b*), or it may appear as a streak of vapor, broadest at the middle, and tapering towards either end (*c*). A third variety consists of *small rows of clouds,* parallel to one another; each successive row becoming shorter, from the widest part of the cloud to the extremities, (*d.*) The name of cirro-stratus is also given to that thin, gauze-like cloud, which sometimes overspreads the whole sky, and through which the sun and moon are dimly visible.

246. CIRRO-CUMULUS. It not unfrequently happens, that the heavens appear as if sown with *little round masses* of clouds, lying near to each other, but perfectly separated by intervals of sky (figure 14., *a*). This cloud is the *cirro-cumulus,* and often arises from a change in the cirrus and cirro-stratus; the bars of the latter being divided across the direction of their length, and the different parts rounding into the cirro-cumulus. Sometimes the reverse occurs, and the cirro-cumulus is seen changing into the cirrus and cirro-stratus.

247. The structure of the cirro-cumulus is not always the same: at one time the masses are very dense and well-rounded; at another their form is irregular, and the sky often presents a *curdled* appearance, when covered with this cloud. Sometimes the cirro-cumulus is so light and fleecy, that the rays of the sun, as they traverse it, are scarcely dimmed. Humboldt found them

---

Describe the cirro-stratus.
How is it produced?
What are some of its varieties?
Describe the cirro-cumulus.
Whence does it arise?
State some of the peculiarities of this cloud

even so delicate that he was able to discern through them the spots on the moon. The last two classes of clouds, like the cirrus, float at a very lofty height.

Fig. 14.

CIRRO-STRATUS (b, c, d), CIRRO-CUMULUS (a), NIMBUS (e), CUMULO-STRATUS (f).

218. CUMULO-STRATUS. The variety of cloud to which this name is given, combines the characteristics of the *cumulus* and *stratus*. Its base consists of a horizontal *stratum* or layer of vapor, from which rise large, overhanging masses of *cumulus* (figure 14., f). Some-

---

What is said respecting the height of the cirro-stratus and cirro-cumulus?
Describe the cumulo-stratus.
Of what does it consist?

times contiguous cumulus clouds unite, and passing into the state of cumulo-stratus, form groups of immense size. This cloud is seen in perfection upon the approach of a thunder-storm, when the cumulus clouds, driven together by the wind, are piled upon each other, and assume those peculiar forms commonly known by the name of *thunderheads*.

249. This modification also frequently arises, when the cumulus is pierced by the cirro-stratus; and it is by no means unusual to see *four or five* parallel bars of the cirro-stratus, one above the other, passing through the same pile of clouds, which then present successive tiers of the cumulo-stratus.

250. NIMBUS, OR RAIN-CLOUD. This cloud is so called from the Latin word, *nimbus, a rainy dark* cloud; it possesses no peculiarity of form, but is distinguished by its uniform gray tint and fringed edges (figure 14., *e*). It is usually composed of some of the preceding classes of clouds, so blended together that they cannot be distinguished, and is produced by a change in their structure, the result of an increase in density.

251. The nimbus often originates in the cumulo-stratus, which, as it increases in thickness, frequently assumes a *black* or *bluish* tint. In a short time this hue changes to *gray*, a circumstance which indicates that the nimbus is formed and rain descending.

When is this cloud most perfectly formed?
Under what other circumstances is the cumulo-stratus seen?
Describe the nimbus.
How is it distinguished?
Of what does it consist?
How is it caused?
In what cloud does it often originate?
What does a gray tint indicate?

CHAPTER IV.

OF DEW.

252. DEW *is the moisture spontaneously deposited upon the surfaces of bodies exposed to the atmosphere, when the latter is free from the presence of fogs and rain.*

253. The whole subject of dew was most happily illustrated by the observations and experiments of Dr. Wells, in 1812; and the theory which he then advanced is now generally received, supported as it is by a vast assemblage of facts.

254. DEPOSITION. The deposition of dew is caused by *the unequal radiation of heat from the atmosphere and the substance bedewed.* During the day, the bodies, that either compose the solid crust of the earth or clothe its surface, become heated by the solar rays, and the lower stratum of that portion of the atmosphere which is directly above, is then likewise raised in temperature, and its capacity for moisture increased.

When, however, the night comes on, and even before, the earth and air, radiating their acquired heat into free space, sink in temperature; but the loss of the former is greater than that of the latter, since, during the night, as experiments show, the air a few feet above the ground, is sometimes warmer than the surface of the soil, by *fifteen* degrees.

It thus occurs, that the stratum of air immediately in contact with the earth is cooled down by the latter, beyond the point of saturation; and the excess of humidity is condensed, upon the substances that form its surface, in drops of dew. (Art. 65.)

255. It may therefore be assumed as a principle, *that dew never begins to be deposited upon the surface of*

What is the subject of chapter fourth? Define dew.
Whose theory is generally received?
How is the deposition of dew caused? Explain the process.
How much warmer is the air sometimes than the ground?
What principle may be assumed?

INFLUENCE OF THE ATMOSPHERE. 107

any body, *until it is colder than the contiguous atmosphere;* and, other circumstances being the same, the greater *this difference in temperature,* the greater the *amount of dew.*

The quantity of dew deposited within any given time, depends, chiefly, upon the *humidity, serenity,* and *tranquillity* of the atmosphere; and the *constitution, form, surface,* and *location* of the bodies receiving the moisture.

INFLUENCE OF THE CONDITION OF THE ATMOSPHERE.

256. HUMIDITY. That the quantity of latent vapor in the air must regulate, in a great measure, the amount of dew, is perfectly clear, since the latter is nothing else than *condensed atmospheric vapor.*

257. SERENITY. Every thing that favors radiation from the earth, and consequently produces an increase of cold, contributes to the formation of dew. Thus it is copiously deposited on *serene nights;* for the radiation from the earth then proceeds unchecked: while, on the contrary, little or no dew is seen after a *cloudy night;* since the canopy of the clouds reflects back to the earth the heat that is proceeding from it, maintaining its surface and the contiguous air at nearly the same temperature.

If, however, the clouds *separate* only for a *few moments,* the heat escapes from the earth through the intervals, and *dew* is rapidly *deposited;* but if the sky is again suddenly *overcast,* the radiation is arrested, and the heat reflected back to the earth, raising the temperature at its surface, and speedily *evaporating the dew just formed.*

258. These singular changes in temperature were observed by Dr. Wells. On one night, the sky being *clear,* the temperature of the grass, at half past nine, was 32° Fah.; in *twenty minutes* afterwards, the heavens being suddenly *overcast,* it rose to 39° Fah.; in

---

What circumstances influence the quantity of dew?
What is the effect of humidity? What of serenity?
What is the influence of clouds? Give instances.

twenty minutes more, under a *serene* sky, it sunk again to $32°$ Fah. It was also found, that a thermometer lying upon the grass, would rise *several degrees*, if the sky directly above it was covered by a *cloud* only for a few minutes. The influence of clouds upon the *temperature* of the *air* is by no means as great; for while, on one evening, when the sky was obscured for the space of forty-five minutes, a thermometer placed upon the *turf* rose *fifteen* degrees, another, suspended in the *atmosphere* just above, rose but *three and a half* degrees.

259. TRANQUILLITY. In a *calm* night, other circumstances being the same, more dew will be deposited than when it is *windy;* for the wind will not suffer any one portion of air to remain long enough in contact with the cold surface of any body to condense much of its moisture, hurrying it away before it is sufficiently cooled down for this purpose.

260. A *slight* agitation of the atmosphere, however, is of advantage; since, after one portion of air has deposited upon any surface its exuberant moisture, it removes it from the spot, bringing up a fresh portion to the same place, and so on successively; giving time to each to sink to the temperature of the surface bedewed. As the night advances, and the earth becomes still colder, the *same volumes of air*, renewing their contact with the *same surface*, may be again surcharged with humidity, and deposit more dew.

261. EVENING AND MORNING. Dew is often formed towards the *close* of the *afternoon*, in consequence of the earth then losing more heat by radiation than it receives from the slanting rays of the descending sun. It also frequently forms in shady places *just after sunrise;* for the surface of the globe, which has been gradually sinking in temperature during the night, is not

---

Which is most affected by clouds, the air or the ground? Illustrate.
Why does a wind lessen the amount of dew?
What is the effect of a slight agitation of the air?
Why does dew begin to form towards the close of day?
Where does it form after sunrise?

immediately influenced by the warm beams of the sun. Indeed, at this time, more dew is deposited than at any other equal period in the twenty-four hours.

INFLUENCE OF THE SUBSTANCE BEDEWED.

262. CONSTITUTION. Since the production of dew requires that the body bedewed must be colder than the surrounding atmosphere, *all substances, which rapidly lose their own heat and slowly acquire that of others, are susceptible of being copiously bedewed.* On the contrary, substances possessing the opposite qualities contract but little dew.

Under the first class may be included *glass, silk, down, wool,* and, in general, all bodies of a porous texture; while *metals* and *rocks* belong to the second division.

263. If similar plates of polished glass and metal are exposed alike upon the soil during a favorable night, in the morning the glass will be *drenched* with *dew*, but the brightness of the metal will be *scarcely dimmed.* These different results arise from the fact, that, while the glass is deprived by radiation of *ninety hundredths* of its original heat, *twelve hundredths* is all that the metal loses. Besides, the glass, being a *bad conductor*, draws but little warmth from the earth to supply its loss; while the metal, being a *good conductor*, can easily restore any reduction of heat from the warm soil immediately below.

Large plates of metal, exposed throughout the night, never acquire a temperature more than *three or four degrees* below that of the atmosphere.

264. SURFACE AND FORM. A *polished* surface does not radiate so well as one that is *rough* and *uneven;* and the latter is always found, under a like exposure, to receive more dew. *Points* radiate heat most rapidly,

---

Account for its deposition at this time.
What substances are capable of being copiously bedewed?
What not? Give examples.
Account for the unequal deposition of dew on glass and metal.
What is said of polished and rough surfaces in this particular?

110 AQUEOUS PHENOMENA.

and are therefore the coldest portions of a radiating body; hence, of all the globules of dew that form upon blades of grass, the largest are found at the very extremities.

Grass is well known to be copiously bedewed; its form, as just mentioned, causes it to lose its own warmth with great rapidity, while its porous texture prevents it, at the same time, from replenishing its loss from the soil.

265. LOCATION. If a body is *screened* from the open sky, it contracts less dew than when fully *exposed;* for the screen arrests radiation in the manner of clouds; and the difference in temperature between the sheltered body and the contiguous air, is less than that which would exist between the same body and the surrounding atmosphere, were the substance bedewed entirely unsheltered. This has been proved by the experiments of Dr. Wells.

266. In four trials, made with two parcels of wool, in all respects alike, the first of which was placed upon the *upper* side of a board, four feet from the ground, and the second loosely attached to the *under* side, the gain, in dew, was as follows:

	1st night.	2d.	3d.	4th.
	grs.	grs.	grs.	grs.
1st parcel,	14	19	11	20
2d do.	4	6	2	4

We hence perceive, why, beneath the shelter of trees, and on the under surfaces of leaves, but little dew is found.

267. Dew has never been found upon the *surface of large bodies of water;* for whenever the aqueous particles at the surface are cooled, they become heavier than those below them, and sink; while warmer and lighter particles rise to the top. These, in their turn, become

---

What of points?
Why are the largest beads of dew upon the end of the blades of grass?
Why does an exposed body contract more dew than one which is sheltered? Give the results of Dr. Wells' experiments.
Why are the surfaces of large bodies of water free from dew?

heavier and descend; and the process continuing throughout the night, maintains the surface of the water and the air at nearly the same temperature.

Dr. Wells ascertained, by experiment, that even a small quantity of water gains no weight by exposure during a single night.

It appears, from the narrative of the U. S. Exploring Expedition, and from other sources, that on the ocean heavy deposits of dew sometimes occur upon the *decks* of vessels.

268. The exposed parts of the human body are never covered with dew; since the vital heat, varying from 96° to 98° Fah., effectually prevents such a loss of warmth as is necessary to its production.

269. COLOR. A few experiments were made by Dr. Wells, in order to ascertain the effect of color upon dew; but without any decisive results. In 1833, Dr. Stark, of Edinburg, made two experiments, from which he inferred, that under like exposures, *more* dew was deposited upon *dark*-colored bodies, than upon *light*-colored. But the author of this work, from an investigation prosecuted by himself during the summer of 1846, arrived at the conclusion, that *color exerts no influence whatever upon the quantity of dew*. This fact might also be inferred from the experiments of Dr. Bache on heat, which clearly show, that the amount of radiation is not affected by color.

270. OBSERVATIONS. The observations, which have been made in various regions of the globe, in regard to the occurrence of dew, strongly corroborate the theory of Dr Wells. In Bengal, during the month of November, the nights are beautifully *serene*, and accompanied with *heavy* dews; but in April and May, when *high winds* prevail, with a *close* and *cloudy* atmosphere, *no*

What experiment was made by Dr. Wells?
What is stated in the narrative of the Exploring Expedition?
Why is dew never found upon the human body?
What is said as to the influence of color?
What do the observations made in different regions attest?
Give instances.

dew is deposited. From September to March, the sun glows like an orb of fire over Southern Guinea; but the nights are *cool*, and the parched soil is refreshed with abundant dews. In Egypt, *profuse dews*, like rain, occur in the summer, when the nights are *resplendent with stars;* while at Thebes, where the sky is constantly *serene, abundant dews* are the only moisture that the earth receives from above, during the lapse of many years.

271. FACTS EXPLAINED. The explanation of several familiar facts, depends upon the foregoing principles. Thus, for instance, if, in a warm summer's day, a glass is filled with cold water, the exterior surface is seen covered with moisture; for the reason, that the glass, being *colder* than the *air* in contact, cools the latter below the dew-point. In summer, caves and cellars are damp; because the warm air that enters them from without is *cooled down*, and its humidity either floats in the atmosphere, or is condensed in beads of dew upon the stones.

272. BENEFICENT DISTRIBUTION. The mode in which the blessing of dew is dispensed to the earth, beautifully exemplifies the benevolence of our Creator.

In the temperate climes, where the frequent interchange of sun and shower preserves the earth from the extremes of heat and moisture, very little dew is needed, and but comparatively little is deposited. The regions however within the tropics are deprived of rain for months, and this destitution is partially supplied by the dews, which *precisely* at these seasons are most abundant.

273. The lake and the river are not visited by dew, for no form of vegetation exists within them that needs its presence. To the naked rock it comes but in scanty measure; for there is nothing here to nourish—save, perhaps, the thorny cactus, which, from its very form and

---

What facts are explained upon the foregoing principles?
What does the distribution of dew exemplify?
Give the various illustrations.

nature, is adapted to its situation; and though springing from the cleft of a rock beneath a tropic sun, or striking its roots in the sands of the desert, is capable of deriving from the air an adequate supply of moisture.

274. Upon the foliage of the grove very little dew is deposited, in consequence of the inclined position of the leaves, their frequent motion, and the shelter they afford each other. Nor is it needed; for the natural supply of moisture rises deep from the soil through the parent trunk, diffusing itself throughout every branch to the remotest extremity of the slenderest bough.

275. The dew, however, blesses, in all its invigorating exuberance, the humble plant and tender herbage, a vast class of vegetable life, at once the most perishable and the most useful; it is the first of all to feel the effects of drought, and yet it is that which is necessary to the very existence of man. From the *field*, not from the forest, comes our support; and the failure of a single plant, the grass or the bladed grain, may bring upon a nation scarcity and famine.

CHAPTER V.

OF HOAR-FROST AND SNOW.

276. HOAR-FROST. Hoar-frost is produced in the same manner as dew. Late in the spring, and early in the fall, the surface of the earth, during serene nights, sinks in temperature *below the freezing point*, while the atmosphere, a few feet above, is *higher by several degrees.*

The moisture which is then deposited becomes congealed in sparkling crystals; and the stems of plants and the branches of low shrubs are often adorned with fringes, formed of the most beautiful and delicate crystallizations.

What is the subject of chapter fifth?
How is hoar-frost produced?
Describe its appearance.

277. A species of hoar-frost occurs when a warm south wind succeeds a continuance of cold weather. Stone columns and buildings are then covered with a snowy incrustation, composed of an assemblage of minute crystals, caused by the influence of the low temperature of the stone upon the condensed vapor of the air.

The effect of a cold body upon moist air is well shown by the following facts related by Ballantyne, who resided for two years at York Factory, in the vicinity of Hudson's Bay. After narrating the adventures of a hunting expedition in the depth of winter, he thus describes an incident that occurred upon the return of himself and his companions to their dwelling. "It was curious to observe the change that took place in the appearance of our guns after we entered the warm room. The barrels and every bit of metal upon them instantly became white, like ground glass. This phenomenon was caused by the condensation and freezing of the moist atmosphere of the room upon the cold iron. Any piece of metal, when brought suddenly out of such intense cold into a warm room, will in this way become covered with a pure white coating of hoar-frost. It does not remain long in this state, however, as the warmth of the room soon heats the metal and melts the ice. Thus, in about ten minutes our guns assumed three different appearances. When we entered the house they were clean, polished, and dry; in five minutes they were as white as snow, and in five more were dripping wet."

278. Every thing that *prevents the radiation of heat, arrests the formation of hoar-frost.* During the chilly nights of spring, plants that are sheltered by trees are less liable to be frozen than those which are fully exposed; and a slight covering of straw, or even of paper, will often afford an effectual protection. Vineyards have frequently been saved from the effects of frost, by enveloping them during the night in a cloud of smoke.

What effect is caused by a warm south wind, after a period of cold weather? Relate the facts related by Ballantyne.
What arrests the formation of hoar-frost?

279. The effect of a screen in checking radiation, and thus preventing frost, has been finely illustrated by the experiments of David Scott, of India. Throughout the whole region of Upper India, ice is artificially procured by placing upon a layer of dry straw, in the bottom of small pits, and fully exposed to the clear sky, broad, shallow earthen pans, filled with water. Such is the radiation during the night, that a thin crust of ice will sometimes form upon the water, when the temperature of the air on a level with the pits is as high as 41° Fah.

On one occasion, Mr. Scott extended a muslin turban across a pit, three feet above the pans. No ice was formed in the vessels immediately *under* it; but, in several that were partially covered, ice appeared upon the part of the water *beyond* the shelter of the muslin; while the surface *beneath* the turban remained in a fluid state. Two strings, crossing each other at a lower height above a pan, under favorable circumstances, divided the ice into four quarters, the water *beneath* the strings continuing *unfrozen*.

SNOW.

280. *Snow is the frozen moisture that descends from the atmosphere when the temperature of the air at the surface of the earth is near or below the freezing point.*

281. SNOW-FLAKE. At moderate heights, and in the temperate regions, snow commonly falls after several days of severe frost when the weather has moderated. The largest flakes occur when the air abounds with vapor and the temperature is about 32° Fah.; but as the moisture diminishes, and the cold increases, the snow becomes finer.

In the former case, it is not unusual to observe flakes *an inch in diameter;* and in the latter, they only measure a few hundredths of an inch.

Illustrate the influence of a screen, by the experiments of Scott.
Define *snow*. When does it usually fall?
Under what circumstances do the largest flakes occur?
Under what circumstances do the smallest? How large are they?

At Bossekop a fall of snow occurred when the thermometer stood at 10° Fah., and the *diameter* of the flakes then scarcely exceeded *seven hundredths of an inch.*

The snow-flake is composed of regular crystals, and its beautiful figures and rich diversity of forms have ever excited the admiration of observers. In solid ice, the crystals are so blended together that their symmetry is lost in the compact mass; but in snow, they are perfectly developed, when the flakes descend through a calm atmosphere. Any agitation of the air, or an increase of moisture or temperature, destroys their delicate structure.

If the crystals of snow were solid, they would be transparent, like other crystallized bodies; but they contain air, and to this circumstance is attributed their brilliant whiteness; for the air preventing the ready transmission of light through the snow-flake, the rays are copiously reflected from the assemblage of crystals

The bulk of snow which has just fallen is ten or twelve times greater than that of the water obtained by melting it.

282. Though single crystals always unite at angles of 30°, 60°, or 120°, they nevertheless form, by their different modes of union, several hundred distinct varieties.

Scoresby, a celebrated Arctic navigator, has enumerated six hundred, and delineated ninety-six; and Kaemtz has observed twenty more, not figured by Scoresby.

283. SNOW-CRYSTALS. Although the varieties are so numerous, they are all comprised under *five* principal classes.

1st. Crystals in the form of *thin plates;* they are generally very thin, transparent, and of a delicate

---

How small are they?
Of what is the snow-flake composed?
How is the whiteness of snow caused?
What is said of the bulk of snow?
State the number of varieties of snow-crystals.
In how many classes are they comprised? Describe them.

structure. This class includes many remarkable varieties, which are represented by the first twenty-five figures of the annexed cuts, (15., 16.)

2d. Flakes either possessing a *spherical nucleus*, or a *plane figure*, studded with needle-shaped crystals, (figure 26.)

3d. Slender *prismatic* crystals; usually *six-sided*, but sometimes having only *three sides*.

4th. Pyramids with *six sides*; (figure 27.)

5th. Prismatic crystals, having, perpendicular to their length, both at the *extremities* and in the *middle, thin, six-sided plates*; (figures 28., 29. and 30.) The last

Fig. 15.

SNOW-CRYSTALS.

Fig. 16.

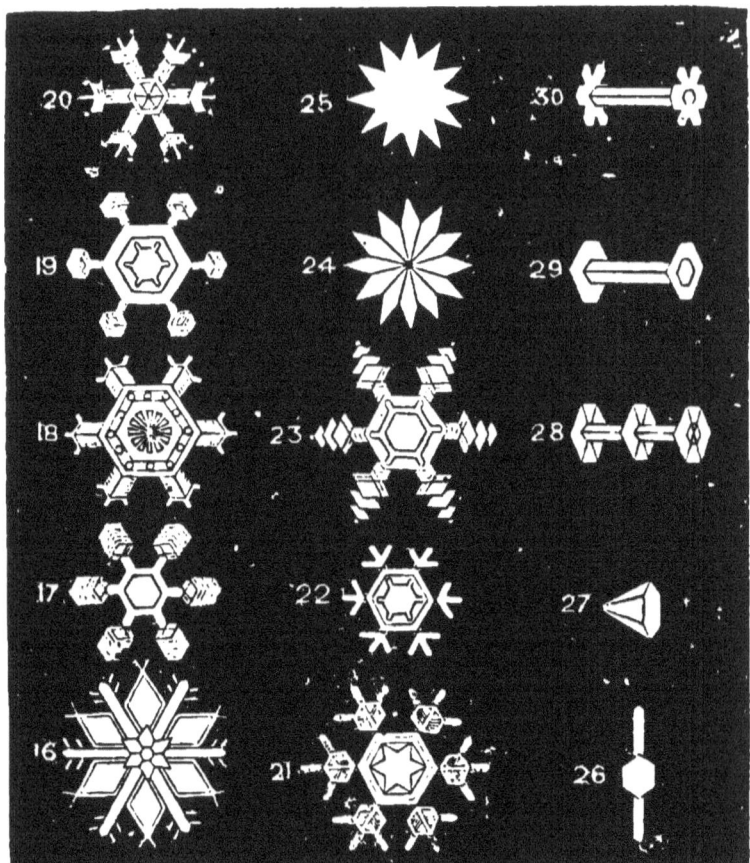

SNOW-CRYSTALS.

two classes are extremely rare, Scoresby having observed the *fourth* but *once*, and the *fifth* only *twice*, in all his voyages.

Flakes belonging to two consecutive falls of snow, possess different figures; but those which descend during the same storm, are usually alike in this particular.

284. NATURAL SNOW-BALLS. Balls of snow are sometimes formed by the action of a high wind upon light snow. Prof. Cleaveland, of Brunswick, in Maine,

---

What is said of the crystals that fall during the same storm?

observed, on the first of April, 1815, a great number of snow-balls scattered over the fields, varying from one to fifteen inches in diameter. They had evidently been caused by the wind rolling up the snow, as the track of the balls was distinctly visible. In 1830, similar balls were seen by Mr. Sheriff, in East Lothian, scattered over a wide extent; some of the masses being eighteen inches in diameter.

285. But the most remarkable exhibition of this kind was beheld by Mr. Clarke, of Morris county, New Jersey, in January, 1808. A crust having formed upon the snow that had previously fallen, a light snow soon after occurred, covering the glassy surface to the depth of three-quarters of an inch; the sky then suddenly became serene, and a high wind arose. Beneath the force of the gale, small portions of snow would slide along for the distance of ten or twelve inches, when they would begin to revolve, rapidly increasing both in length and diameter. Where the descent of the ground favored their formation, masses rolled up to the *size of a barrel*, and, as far as the eye could see, the dazzling surface was covered with balls and cylinders of snow; varying in magnitude from *ten inches to three feet in diameter*. Upon examination they were found to be hollow at each end, almost to the centre, and as round as if they had been so many logs of wood turned in a lathe. The cylinders covered nearly 400 *acres*, and their number was judged to be nearly 40,000.

286. RED SNOW. In 1819, Capt. Ross beheld snow of a brilliant *crimson* hue, clothing the sides of the mountains at Baffin's Bay; rising, according to his report, to the height of *several hundred feet*, and extending to the distance of *eight miles*.

Snow of this tint is not, however, confined to the Arctic regions. Raymond had previously observed it in the Pyrenees. In 1818, vast masses were spread over the Italian Alps and Apennines, and five years before,

---

Relate the several accounts of *natural snow balls*.
What is said of *red* snow?

the whole range of the last-mentioned chain was covered with rose-colored snow. The same phenomenon was seen by Scoresby, Parry and Franklin, in high northern latitudes, and the navigators of the southern hemisphere have found red snow in great quantities at New Shetland, 62° S. Lat.

287. In snows of great depth, the accounts differ in regard to the thickness of the colored stratum. Ross conjectured, that, in the Arctic mountains, the crimson hue penetrated to the depth of several feet below the surface; while others could not detect its existence beyond one or two inches.

Among the Alps, the red snow is usually discovered in little sheltered hollows, in layers not exceeding two or three inches in thickness: though these are sometimes situated far beneath the general surface of the snow.

288. GREEN SNOW. When the French meteorologists, Martin and Bravais, traversed a field of snow at Spitzbergen, in 1838, it appeared of a *green hue*, wherever it was pressed by the foot. The coloring matter seemed to reside just below the surface, which was brilliantly white.

Upon another excursion, the first observer beheld the green particles spread like dust over the snow, which was also tinted green beneath the surface, and upon the sides of the field.

289. CAUSE. These singular hues are produced by the presence of an infinite number of a certain class of microscopic plants, which from their great tenacity of life, are capable, not only of existing at a very low temperature, but even of flourishing with extraordinary vigor.

These minute vegetable forms are composed of globules, which vary in diameter from *one-thousandth* of an inch to *one three-thousandth*. Each globule is divided into seven or eight cells, filled with a liquid, in which

---

What of *green* snow?
To what cause are those colors attributed?

live a great number of *animalcules.* The cells are generally red, which is supposed to be their original color, the green tint being probably acquired by exposure to the air and light.

These extraordinary hues may, therefore, be regarded as originating in the same plant, in different stages of development.

290. USES OF SNOW. Snow subserves many important purposes. Gathered in exhaustless stores upon the high mountains of the globe, it feeds, as it gradually melts beneath the heat of summer, thousands of rivers, which, flowing on from clime to clime, enrich the soil and crown the land with plenty.

The snow-capped mountains are the natural refrigerators of the glowing regions that lie within the tropics; cooling the winds that pass over them, and mitigating the fierce temperature of the atmosphere.

In the higher latitudes, where the winters are severe, the snow forms a warm covering for the soil, and defends vegetation from the rigors of the frost. It is well known, that grain, during an open winter, is frequently destroyed by the cold; and, in the mild climate of England, Alpine plants have perished, in consequence of being deprived of their natural covering of snow.

During the long night of the polar climes, the intensity of the darkness is diminished by the presence of the snow; inasmuch as it *reflects,* instead of *absorbing,* like the bare ground, the faint light that there proceeds from the sky. Nor is it to be forgotten, that, in these inclement regions, the wretched natives would be unsheltered during the winter, were it not for the snow; since this, when cut into blocks, supplies the Esquimaux with the means of constructing their huts.

What is said in regard to the *uses of snow?*

CHAPTER VI.

OF HAIL.

291. HAIL. *The ice that descends in showers, and usually in summer, is called hail.* It is different from sleet, which is nothing more than frozen rain, and occurs only in cold weather.

292. STRUCTURE. Hailstones are generally *pear-shaped*, and if they are divided through the centre, they are found to be composed of alternate layers of ice and snow, around a white, snowy nucleus, resembling the coats of an onion. The surface is rough, and is sometimes studded with icicles.

293. SIZE. Hail varies in size, from stones as small as a pea to those which are several inches in circumference. Benvenuto Cellini relates in his memoirs, that during his journey from Italy to France, he was overtaken by a terrible storm in the vicinity of Lyons; hailstones falling of the *size of lemons*, and with sufficient force to kill even men and cattle.

At Roncesvalles, in August, 1813, there fell upon a division of the British army a storm of hail, in which the stones ranged in size from a bean to a *hen's egg*. The tin camp-kettles of the soldiers were indented by the masses of ice, some of which were round, and armed with icicles three inches in length.

In May, 1847, hailstones of immense size descended near the town of McDonough, in Georgia; one of them was measured an hour after it fell and found to be *ten inches in circumference*. During a terrific storm, that occurred at Cazorta, in Spain, on the 13th of June, 1829, the roofs of the houses were broken in by the hail. Some of the stones are stated to have weighed nearly *four pounds and a half.* It is probable that such extra-

---

Define hail.
What is the form and structure of *the hailstone*?
What is said of its *size*?
Narrate the facts stated.

ordinary masses as those which have been mentioned, are formed by the union of several hailstones frozen together.

GEOGRAPHICAL DISTRIBUTION.

294. Hailstorms are *most frequent in the temperate climes*, and *rarely* occur within the tropics, except in the vicinity of mountains whose summits tower above the limit of perpetual frost. Although by no means common, they are not unknown in the high northern latitudes. Simpson, on the 12th of August, 1839, was exposed in the straits of Boothia, in 68° 32′ N. Lat., to a tremendous thunder-storm, accompanied with torrents of rain and heavy showers of *hail*.

It is mostly in *summer*, and usually at the *hottest part* of the day, that hail is observed to fall. Scarcely any occurs in the night.

ORIGIN.

295. The structure of the hailstone shows that it is not formed *at once;* for the concentric layers around the snowy nucleus, consist of different accessions of moisture, *successively* condensed and congealed upon the surface of the stone.

The light, porous texture of the snowy centre, seems to indicate, that the place of origin must be some region in the atmosphere where the *air* is *rare*, and the cold *intense;* since the appearance of the centre is similar to that presented by a drop of water, when frozen under the exhausted receiver of an air-pump.

296. It is necessary then for the production of hail, that a warm, humid body of air should mingle with another so extremely cold, that their temperature, after uniting, shall be *below the freezing point*. This combination must also take place during the *warmest* period

---

Where do hailstorms frequently occur?
Where rarely?
When do they usually prevail?
What indicates that the hailstone is not formed at once?
Where must it originate?
What conditions are necessary for the production of hail?

of the *year* and the *day*. In accounting for an intense degree of cold under such circumstances, consists the great difficulty of explaining the phenomena of hailstorms.

297. Until within a few years, almost every meteorologist attributed the cold of hailstorms to the *agency of electricity*. It is well known that air, when electrified, is expanded, and that expansion produces cold. It was therefore imagined, that the electrified state of the atmosphere before a storm, caused such a reduction of temperature as to freeze the falling moisture and produce hail.

Volta, a distinguished philosopher of France, supposed the cold to be the result of *evaporation*, but employed electricity in a singular manner, as explained below.

298. VOLTA'S THEORY. According to this theory, two clouds, differently electrified, are supposed to extend through the sky, one directly above the other. The cold, caused by evaporation from the upper surface of the lower cloud, is considered to be so intense, that the vapor is frozen, and the nucleus of the hailstone then formed. Its size is afterwards increased by the humidity it gathers in vibrating backwards and forwards between the two clouds, like the dancing figures upon electrical plates. (C. 969.) At last it becomes so large, as to break through the lower cloud, and fall to the earth.

299. The sanction of a great name gave weight to this fanciful view, and in 1821, throughout the southern districts of France, which are peculiarly liable to hail storms, hail-rods were erected, in order to draw the electricity from the clouds, and thus protect the vineyards. Their efficacy, however, is exceedingly questionable.

The electric hypothesis is, moreover, at variance with facts. The forests, which constitute a vast assemblage of hail-rods, are often ravaged by hail; and in the tor-

---

What effect has been attributed to electricity?
Explain Volta's theory.

rid zone, where the development of atmospherical electricity is greatest, hailstorms are almost unknown.

300. OLMSTED'S THEORY. Prof. Olmsted, of Yale College, considers *electricity* as an *effect*, and not the *cause* of hailstorms. According to his theory, which has been very extensively received, the *cold body* of air derives its low temperature, *not from electricity*, but from some *known source of cold;* and the combination, which occasions the hail, may arise in various ways, the principal of which appear to be the following.

301. First. *An exceedingly cold wind, coming from a region far above the limit of perpetual frost, may meet with a current of warm air, blowing from a point many thousand feet below this limit.*

Secondly. *By the force of whirlwinds, large volumes of warm air from the surface of the earth may be suddenly transported into the higher and colder regions of the atmosphere.*

Thirdly. *In the vicinity of lofty mountains, cold blasts are frequently known to sweep down their sides from the snowy peaks and glaciers, and mingle with the warm atmosphere of the vales.*

Each of these methods we will discuss separately.

302. CURVE OF PERPETUAL CONGELATION. In Art. 53, we have seen that a point can be reached in every latitude, where moisture, once frozen, always remains so. An imaginary line passing through these points, and extending from pole to pole, forms what is termed the *curve of perpetual congelation*, which possesses the peculiar figure shown in the annexed cut.

Does the electric theory agree with facts?
What are Professor Olmsted's views in regard to electricity?
Whence comes, according to his theory, the cold of the hailstorm
In what three ways may hailstorms arise?
What is the curve of congelation?

126   AQUEOUS PHENOMENA.

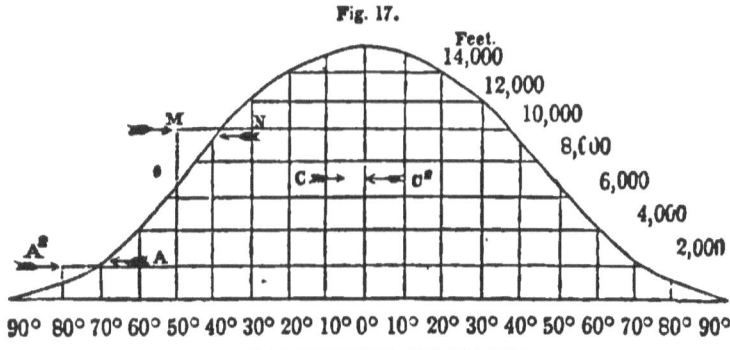

Fig. 17.

CURVE OF PERPETUAL CONGELATION.

303. The heights of the curve from the surface of the globe vary but little from the equator to Lat. 30°; but from 30° to 60° the change is very great, and the line rapidly approaches the earth.

The difference in the height of the points of congelation, for every five degrees of latitude, is presented in the following table:

Lat.	Difference of height in feet.
0° to 5°	122
5° to 10°	388
10° to 15°	569
15° to 20°	779
20° to 25°	689
25° to 30°	1,438
30° to 35°	928
35° to 40°	1,648
40° to 45°	1,358
45° to 50°	1,398
50° to 55°	1,348
55° to 60°	1,228
60° to 65°	1,168
65° to 70°	959
70° to 75°	809
75° to 80°	628

304. ACTION OF OPPOSITE CURRENTS. We are now to imagine, for the sake of illustration, that a north

---

Describe its peculiarities.

wind, originating in 50° N. Lat., moves horizontally at the rate of *sixty miles per hour*, at an altitude of *ten thousand feet ;* while a south wind blows simultaneously from 30° N. Lat. with the like velocity, and at the same height.

If they are upon the same meridian, they will meet in ten hours at 40° N. Lat., and since the first wind commences its course at M, *three thousand feet above* the limit of constant frost, it will be extremely *cold ;* while the south wind will be comparatively *warm*, as it proceeds from a region, N, *two thousand feet below* the boundary of perpetual congelation.

By the union of air, thus widely differing in temperature, the inherent atmospheric vapor is both condensed and frozen, and the central portion of the hailstone formed, which, in its descent to the earth, is gradually enlarged by constant accretions of frozen moisture.

305. The prevalence of such opposite currents as have just been supposed, has already been shown (Art. 222); and it is by no means improbable that, in their ceaseless circuits, there are times in which they encounter each other. It may be asked, how can the different winds preserve their respective temperatures, in traversing a distance of ten degrees? To this it is answered, *that a fluid in motion can pass through a fluid of the same kind in repose, and differing in respect to heat, without suddenly changing its own temperature.* The waters of the Gulf-stream, flowing through the North Atlantic from the torrid zone, are warmer than the ocean, even in the latitude of Newfoundland.

306. The occurrence of hailstorms, under these circumstances, substantially agrees with facts. It will be seen, by referring to the figure, that the mingling of opposite winds, at a lofty elevation, in *the tropics*, C, C^2, would occasion nothing but a combination of *warm cur-*

Explain the action of opposite currents.
Why can the currents preserve their respective temperatures ?
Show to what extent the occurrence of hailstorms, under these circumstances, accords with facts.

rents, and in the *polar climes* of *cold currents*, A, A²; in neither case could hail be the result of the union.

In the temperate regions, the admixture of warm and intensely cold currents can only be found, and precisely within these limits hailstorms are most prevalent.

Their frequency in summer is attributed to the circumstance, that the opposing winds are then, most easily set in motion by the increased energy of the solar rays.

307. The space ravaged by hailstorms, often indicates the presence of aërial currents, the devastations being frequently confined to a *long* and *narrow* strip of country. Sometimes the storm proceeds in two parallel tracks, leaving the intervening region entirely uninjured.

Thus a hailstorm once commenced in the south of France in the morning, and in a few hours reached Holland. The places desolated formed two parallel paths from S. W. to N. E.; the length of one was 435 miles; and that of the other 497 miles. The average width of the eastern track was five miles, and that of the western ten; and upon the space comprised between them, which was twelve miles and a half in breadth, no hail fell, but only a heavy rain.

308. ACTION OF WHIRLWINDS. It has been stated, (Art. 132,) that whirlwinds are not always vertical, but frequently *inclined* towards the earth. In consequence of this position, the gyratory motion of the whirl (if its diameter is considerable) will, doubtless, often carry up hot and humid air from the surface of the earth into the higher regions of the atmosphere, bringing down in return large volumes of cold air from the upper strata; thus causing such a combination as results in the production of hail. This action will be more extensive and energetic if, as some suppose, whirlwinds at times exist whose axes are *parallel* to the horizon.

309. It must also be remembered, that in the *vortex* of the whirlwind the air is *rarefied*, and into this partial

---

Explain the action of whirlwinds.

void the cold air from above will descend, by reason of its superior weight; while below, on account of the pressure of the surrounding atmosphere, warm currents will stream under the base into the vortex. Here, then, may evidently occur a union of hot and cold air, differing so greatly in temperature that the condensed moisture will freeze into hail.

The cold, arising from the *rarefaction* of the air in the centre of the whirlwind, also contributes to the formation of hail.

310. INFLUENCE OF HIGH MOUNTAINS. In the vicinity of those lofty mountains, whose peaks are always covered with ice and snow, destructive hailstorms frequently occur. The south of France, which lies between the Alps and Pyrenees, is annually ravaged by hail; so great is the ruin to the productions of the soil, and especially the vineyards, that the yearly loss to the national revenue was estimated, by the Linnean Society of Paris, at fifty millions of francs, or *nine millions three hundred and seventy-five thousand dollars.*

In Peru, hail has been seen to fall; and on the 17th of August, 1830, it covered the streets of Mexico to the depth of *several inches.*

311. That such phenomena should arise in these and similar localities, is by no means surprising: for cold blasts of wind descending from the snowy summits of the neighboring mountains, and mingling with the warm air of the plains, could doubtless occasion these results; and the existence of such breezes is fully established.

312. HAIL IN SOUTHERN INDIA. Hail sometimes occurs within the tropics, even at a distance from those mountain-chains that rise above the limit of perpetual frost. Thus in India, in 16° 30′ N. Lat., during the year 1825, hailstones fell at Darwar, of the size of pigeons' eggs; and in a similar storm, which happened at

---

In what localities do hailstorms occur?
Give instances.
What is the cause of hail in these regions?
Does hail ever occur at a distance from snow-capped mountains?

Trinconopoly, in 1805, the stones were as large as walnuts.

313. In conclusion, we may say in regard to this subject, that at present it is not fully understood. Much valuable information has been gathered, but hitherto no theory has been advanced, which completely accounts for all the facts that arise.

Give instances.
Is this subject fully understood?

PART IV.
ELECTRICAL PHENOMENA.

CHAPTER I.

OF ATMOSPHERIC ELECTRICITY.

314. THE atmosphere is usually electrified. The means employed for collecting its electricity differ according to the object proposed; for we may desire to conduct our investigations at one time in the *lower* regions of the atmosphere, at another in the *higher;* or the air may be explored to a great distance *horizontally*.

315 ELECTROMETERS. For ascertaining the electric state of the atmosphere near the surface of the earth, Volta's electrometer is sufficient. An electrometer is an instrument which serves to *indicate* and *measure* electricity. The one just mentioned consists of a glass jar, surmounted by a pointed, metallic rod; and to the lower end of the rod, which enters the jar, two fine straws are loosely attached. The pointed rod, collecting the electricity from the air, the two straws become similarly electrified and recede from each other, (C. 957); the amount of divergence measuring the intensity of the fluid.

316. Insulated rods of iron are erected for testing

What is the subject of part fourth?
What of chapter first?
What is the usual state of the atmosphere?
Why are different means employed for collecting its electricity?
What are electrometers?
How is the electric state of the atmosphere near the earth ascertained?

the air at greater elevations. By means of its pointed summit, the entire conductor becomes charged with atmospheric electricity, the nature of which is easily determined by the electrometer.

At the Kew Observatory, near London, the conductor is a conical tube of thin copper, raised sixteen feet above the roof; to the top of the tube a lamp is affixed, its ascending stream of smoke and heated air being an excellent collector of electricity.

Where a fixed apparatus is not at hand, observations may be made by discharging metallic arrows into the air, in the way hereafter to be described.

317. Experiments are made in the higher regions of the atmosphere by the aid of *kites* and *balloons*. The string of the kite must be wound with fine wire, in order to convey the electric fluid from the sky; and it must also be insulated, by attaching the lower end either to a silken cord or glass pillar. Small, stationary balloons are sometimes employed, the strings of which are arranged and fastened in the same manner.

Occasionally meteorologists ascend in balloons for the purpose of making observations.

318. The method adopted by Mr. Crosse, of Broomfield, near Taunton, for exploring the atmosphere in a horizontal direction, is the following. Upon some of the loftiest trees on his estate, strong poles are firmly fastened, and a copper wire extended from tree to tree; its length was, originally, a *mile and a quarter*, but is now about 1600 feet. The wire, being perfectly insulated, forms a conductor, conveying the electricity of the atmosphere to the room of the observer; where one end of it terminates in an insulated brass ball, near which is a receiving ball, connected with the ground.

In all apparatus for collecting atmospherical electricity, the most careful and certain arrangements should be made for conveying harmlessly to the earth any excess that may accumulate.

How at greater elevations?
For what purpose are kites and balloons employed?
What was the object of Mr. Crosse's apparatus? Describe it.

By the aid of the instruments just described, much important knowledge has been acquired in regard to the electric condition of the atmosphere.

319. ELECTRIC CONDITION OF THE ATMOSPHERE. In the *ordinary state* of the atmosphere, its electricity is invariably *positive;* but when the sky is *overcast*, and the clouds are moving in different directions, it is subject to *great* and *sudden variations; rapidly changing* from *positive* to *negative*, and back again, in the space of a few minutes. Upon the first appearance of *fogs, rain, hail, snow* and *sleet*, the electricity is generally *negative;* it then changes to *positive*, gradually *increasing in strength*, and then *decreasing* in the same manner; the alternations both in the strength and nature of the electricity occurring every three or four minutes. Similar changes are observed upon the approach of a thunder-cloud.

The atmosphere is highly electrical, either when *hot weather succeeds* a series of *wet days*, or *wet weather follows* a series of *dry days*.

320. ANNUAL VARIATION IN INTENSITY. The electricity of the atmosphere is *stronger* in *winter* than in *summer*, and, by comparing observations from month to month, a gradual *decrease* in *intensity* is perceived from January to July, but an *increase* from July to January. During the winter the electricity *strengthens with the cold.*

321. DAILY VARIATION. At *sunrise* the electricity of the air is *weak*, but as the day *advances*, it *increases* in power, until 6 or 7 o'clock, A. M. in summer, 8 or 9 in spring and autumn, and 10 or 12 in winter; it then begins to *diminish*, and by 2 P. M. is hardly stronger than at sunrise. In summer, it continues to decrease till some time between 4 and 6 P. M., and in winter is weakest about 5 P. M.

After this period the electricity again becomes strong-

State the facts in regard to the electric condition of the atmosphere.
What is said respecting the annual variation in intensity?
What of the daily variation?

er, advancing in intensity until about two hours after sunset; when it once more begins to abate, growing more and more feeble until sunrise.

Thus, during the day, there is a regular fluctuation in the strength of the atmospheric electricity; *two periods* occurring when its *intensity is greatest*, and *two* when it is *least*.

322. VARIATION IN ALTITUDE. The electricity of the air *increases in strength* with the *altitude*. This is shown by the following experiment, made by Bequerel and Breschet, at the monastery upon the Great St. Bernard.

Having extended upon the ground a piece of gummed silk, ten feet long and seven wide, the experimenters placed upon it an electrometer; to this they attached one end of a silk cord, into which was twisted a fine wire, the other end of the cord being fastened to an iron arrow. By means of a bow, the arrow was shot upwards to the height of 250 feet; and as in its ascent the electricity of the air was gradually collected and conveyed along the wire to the electrometer, the straws of the latter were seen to diverge more and more, and at length to strike the sides of the glass jar.

When the cord was detached, the electricity of the straws was discovered to be *positive*.

323. In order to determine whether this increased divergence was really caused by the superior energy of the electricity residing in the higher regions of the atmosphere, the arrow was discharged *horizontally* to the same distance as before; but, as it speeded on its course, no increased electric action was manifested by the electrometer.

324. Experiments for the same purpose were made by two celebrated French philosophers, Gay Lussac and Biot, during their aërial voyage in 1804. From the car

---

What is said of variation in altitude?
Relate the experiment of Bequerel and Breschet.
Describe that of Gay Lussac and Biot.
What inference is drawn from both these experiments?

of their balloon was suspended a wire 170 feet long, to the lower end of which a metallic ball was attached; the upper end being connected with an electrometer in the car. By means of this apparatus, these observers were enabled to note the electrical changes occurring in the atmosphere at different heights; and, from their various observations, arrived also at the conclusion, *that the electricity of the atmosphere was positive, and increased in strength with the altitude.*

ORIGIN.

325. EVAPORATION. One of the most abundant sources of atmospherical electricity is *evaporation*. It was shown by Volta, whose experiments were confirmed by those of Saussure, that *electricity was produced when water was evaporated*. But it appears from the late researches of Pouillet, that this is only the case when the water is *not pure*, and *chemical decompositions occur*. If distilled water is evaporated, no electricity is developed; but if a little chalk, lime, salt, or other foreign matter is dissolved in the water, the rising *vapor* is then *positively electrified*, and the *vessel* containing the fluid *negatively*.

326. Now the waters of the earth are generally in this latter condition, being seldom pure, and the *vapors*, which are constantly ascending from the ground, will therefore possess *positive* electricity, and the *earth negative*.

The briny waves of the ocean also contribute their share, and supply the air with a great amount of positive electricity.

327. The process of evaporation advances invisibly and in silence; and, for this reason, we might easily undervalue its agency in accumulating those vast stores

---

What is the first source of atmospheric electricity?
What is said in regard to the experiments of Volta, Saussure and Pouillet?
In consequence of evaporation, does the air become positively or negatively electrified?
Is the amount of electricity thus developed supposed to be great?

of electric matter which arm the storm with such terrific power. But when we reflect, that more than *two hundred millions of hogsheads* of water are computed to rise *daily* in vapor from the Mediterranean, we shall find no difficulty in believing, that this influence is one of the most energetic causes of atmospherical electricity.

328. CONDENSATION. *Condensation*, or the *change which vapor undergoes when returning to a fluid state by a decrease of temperature*, is another fruitful source of electricity. This is shown from the great amount of electricity occasioned by the condensation of steam, as it issues from the boiler of an engine.

In one instance, the steam which rushed from the safety-valve of an insulated locomotive, was found to develop *seven times* the amount of electricity produced by an electrical machine, having a plate of glass *three feet in diameter*, and making *seventy revolutions in a minute*. Machines in which the electricity was generated by steam, have been constructed of such power, that a spark *twenty-two inches long* has been obtained from the prime conductor, (C. 964,) of sufficient energy to *inflame shavings*.

329. VEGETATION. The vegetable kingdom also supplies the air with a great amount of electricity.

Plants during the day exhale oxygen gas; in the night, *carbonic acid gas*—and from the experiments of Pouillet it appears that *positive electricity* rises with the latter when the seeds first sprout, leaving the *earth* in which they are placed *negatively electrified*. The same results probably occur during the life of the plant.

330. COMBUSTION. Combustion is still another source of electricity. When any substance is burning, *positive electricity escapes from it*, while the *substance*

What calculation would lead us to this conclusion?
What is the second source?
What is condensation?
How is it shown that condensation produces electricity? Illustrate.
What is the third source of atmospheric electricity? Explain.
What is the fourth?
During combustion, does the air receive positive electricity or negative?

itself is *negatively electrified;* the atmosphere is therefore the reservoir of all the positive electricity originating in this manner.

331. FRICTION. In accounting for the electricity of the atmosphere, the effect of *friction* is not to be disregarded. If a piece of silk is shaken in the air, it becomes electrified; and it is highly probable, that when masses of air, moving in contrary directions, encounter each other, electricity is developed by the *friction of their surfaces.* Such will be the effect, according to Kaemtz, when the masses differ in respect to *moisture* and *temperature;* the *warmer* then becomes *positively electrified*, and the *colder negatively.*

The action of the wind upon terrestrial objects, as rocks, buildings, trees, and hills, may possibly in like manner produce electricity.

CHAPTER II.

OF THUNDER-STORMS.

332. GENERAL DISTRIBUTION. Thunder-storms *prevail* most in the *torrid zone,* and *decrease in frequency* towards *either pole.*

During a residence of six years in Greenland, 70° N. Lat., Gisecke heard the rolling of thunder but once; and, according to the testimony of the arctic navigators, Scoresby, Parry and others, thunder-storms *rarely* occur between the 70th and 75th degree of north latitude; and *never* beyond the *latter parallel.* As respects *time,* they are more frequent during the *summer months.*

What is the effect of *friction?*
If two bodies of air differ in *temperature,* in what manner will the electricity, developed by their friction, be distributed?
What may possibly be the effect of the friction caused by wind?
Of what does chapter second treat?
How are they distributed in regard to *latitude* and *time?*

The prevalence of these laws will be seen from the observations contained in the following table.

Places.	Latitude.	Period of Observation.	No. of days of thunder in the year.	No. of days of thunder during 6 summer-months.
Buenos Ayres,	34° 30′ S.	7 years.	23	13
Rio Janeiro, .	22° 54′ S.	6 "	51	43
Calcutta, . .	22° 30′ N.	1 year.	60	45
Padua, . . .	45° 15′ N.	4 years.	19	14
Paris, . . .	48° 30′ N.	51 "	14	12
St. Petersburg,	59° 56′ N.	11 "	9	8

333. Thunder-storms are most violent within the torrid zone. Here the play of the lightning is incessant, and the crashing bursts of thunder most terrific; and none but those who have actually witnessed a tropical tempest, can form an idea of its awful power. Occasionally, in the higher latitudes, fierce storms occur, like that which was seen by Simpson in the Straits of Boothia. (Art. 294.)

334. ORIGIN. The thunder-storm is produced in the same manner as the common rain-storm; namely, by the condensation of atmospheric vapor; but it *differs in two respects; first*, in the *rapidity* of this condensation, and secondly, in the *accumulation of electricity* resulting therefrom.

335. We have seen that when *vapor is condensed, electricity is developed* (Art 328): the cloud then in the very process of formation becomes electrified, and to its own electricity is added that which collects upon its surface from the atmosphere; whether derived from evaporation, combustion, vegetation, friction, or any other source.

This condensation must be *copious*, or the electricity would be weak; it must also be *rapid*, else it will es-

---

Repeat the table.
Where are thunder-storms most violent?
How does the thunder-storm differ from the common rain-storm?
Whence is the electricity of the thunder-cloud derived?
Why must the condensation be both *copious* and *rapid*?

cape too fast from the cloud, and never collect in sufficient strength.

336. Thunder-storms are usually attended by a change in the direction of the wind, which accounts for the condensation of atmospheric vapor; indeed, one of the most sublime elements of a storm of this nature, is the conflict and raging of opposing currents.

In the Meteorological Register of Yale College are recorded 116 thunder-storms, which occurred between 1804 and 1823. Of this number, *ninety-nine* were either *preceded* or *followed* by an *alteration* in the *direction* of the *wind;* the change in *fifty* instances being from a *south-westerly* breeze to a *north-westerly*.

Since the air abounds with vapor when its temperature is high, the condensation will be most copious if a loss of heat then suddenly takes place. We therefore easily perceive the reason, why thunder-storms are more frequent in summer than in winter, in low than in high latitudes, and their intensity greatest in the tropic climes.

For the same reason they happen more frequently after mid-day than in the morning.

337. ELECTRICAL STATE OF THUNDER-CLOUDS. Since the air surrounding it is a non-conductor, a single thunder-cloud floats in the atmosphere a vast *insulated conductor* (C. 963); its electricity being spread over the surface of the globules of which it is composed, and there retained by the pressure of the atmosphere.

338. Thunder-clouds may be either positively or negatively electrified; and the observations of Mr. Crosse lead to the conclusion, that at times a cloud of this kind is *complex*, consisting of a series of *concentric* bands or zones, *alternately positive and negative;* the electricity being *weakest* at the *edges* of the cloud, and *strongest* at the *centre*.

How is this condensation effected? What fact is stated in proof?
Explain the cause of the differences that exist in the *frequency* and *violence* of thunder-storms.
What is the electric state of a single thunder-cloud?
State Mr. Crosse's opinion.

Thus in figure 18., which represents a section of such a cloud obliquely seen, P P' P", &c., are *positive* zones, N N' N", &c., *negative*, and the number of dashes show the increase of intensity.

Fig. 18.

339. ELECTRIC ACTION OF THUNDER-CLOUDS. The earth may be regarded as a reservoir of electricity: when, therefore, an electrified cloud floats near its surface, it induces the *opposite* electricity upon the ground *immediately beneath it*.

The cloud may approach so near, that the mutual attraction of the two electricities overcomes the pressure of the atmosphere; a union then occurs, and the lightning, at the same moment, is seen darting between the cloud and the earth, and soon after the rolling of thunder is heard.

340. A similar inductive action arises between the clouds themselves; for, if two clouds differently electrified approach each other, the electricity upon the *nearest opposite surfaces augments in intensity*, and often increases to such a degree that a discharge takes place, the lightning then flashing from cloud to cloud. It may sometimes happen, that the path of least resistance will not be *directly* through the air, but from the first cloud to the earth, and from the earth to the second cloud, and under these circumstances the lightning will take the latter route.

341. RETURN-STROKE. When a highly charged thunder-cloud approaches the earth, it induces, as already stated, the opposite kind of electricity upon the ground below, and repels that of the same kind. Should

---

Describe the electric action of thunder-clouds.
When does a flash occur?
What is the influence of one cloud upon another?
Why does the lightning in passing from cloud to cloud sometimes take the earth in its course? What is the return-stroke?

the cloud be extended, and come within striking distance, either of the earth or of another cloud, a flash at one extremity is often followed by a flash at the other. This is called the *return-stroke*, which sometimes occurs with such violence as to *destroy life*, even at the *distance* of *several miles* from the place of the *first discharge*. The mode of action may be explained by means of the following figure.

Fig. 19.

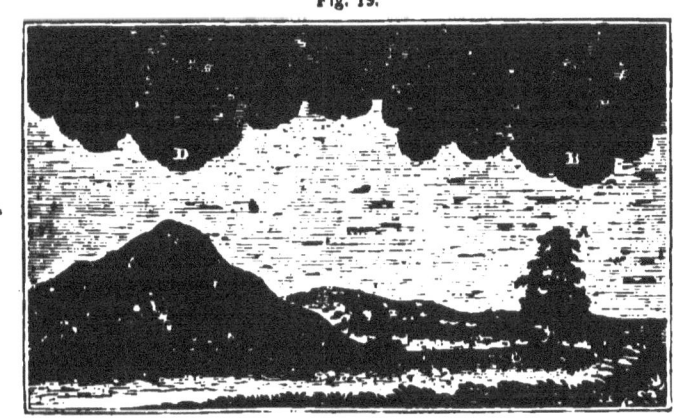

342. Let D B represent a thunder-cloud, *positively* electrified, and within *striking distance* of the tree A; the cloud, at D, being near the summit of the hill, C. By the inductive action of the cloud, the *positive electricity* will be *repelled* from the tree, A, and the summit, C: and both will be highly charged with negative electricity, just before the flash occurs. The moment this happens at B, the cloud becomes *unelectrified*, its inductive action upon C suddenly ceases, the *positive electricity*, which had been *repelled*, instantaneously returns, and, uniting with the *negative electricity* at C, produces an explosion. If, at this time, a person should, unfortunately, be standing upon the top of the hill, his death might ensue.

343. In this manner the following singular facts have

How is it caused?
Illustrate from the figure.
Relate the instances given in Art. 343.

been explained, which happened on the 10th of July, 1785, in the vicinity of Coldstream, in Berwickshire. After a fine morning, clouds were seen in the northwest by the observer Brydone, at about eleven o'clock. Between twelve and one o'clock, the storm being still distant, lightnings were seen darting from cloud to cloud, followed by thunders. Immediately after, Brydone was startled by several loud explosions near his house, like the reports of a gun. At this moment two carts loaded with coals were passing by. The driver and horses of the first were instantly killed, and the coal scattered in all directions, while the driver of the second wagon, which was about twenty yards behind, neither perceived any lightning nor experienced any shock. Upon examination, the hair on the legs and bellies of the horses was found to be singed, and where the wheels rested at the time of the explosions, the tire was melted, and two round holes were discovered in the ground. A quarter of an hour before this event, and at a spot nearly a mile and three-quarters distant, a shepherd of the name of Bell perceived a lamb suddenly fall, while a flame passed before his face. Upon raising the lamb he found it to be dead. A woman, who was cutting grass upon the bank of the Tweed, felt a violent shock upon the soles of her feet, and was thrown to the ground.

During a storm which happened near Manchester, in June, 1835, loud discharges were heard at different points of a road, like the reports of a pistol, and electric flashes distinctly seen; a person is said to have been killed at this time, by an explosion under his right foot.

344. HEIGHT OF THUNDER-STORMS. Though thunder-storms prevail in the lower regions of the atmosphere, they have often been seen at a very great altitude. A storm, observed by Kaemtz, amid the mountains of Switzerland, rose to the height of more than 10,000 feet, and the dwellers in the vale of Chamouni assured him, that storms frequently swept over the

---

What is said respecting the height of thunder-storms?

summit of Mont Blanc. On the peaks of the Cordilleras, a violent thunder-storm was encountered by La Condamine and Boguer, at an elevation of even 16,000 feet. Vitrified rocks have at times been discovered at lofty heights, and as this change is supposed by some to have been effected by lightning, they have sought to determine the altitude of thunder-storms from facts of this kind. The reasoning, however, is inconclusive, for these vitrifications may be owing to other causes, and were we even to grant that they are produced by lightning, the case is by no means proved; since a flash sometimes passes between the *clouds* and the *earth*, when the *former* are *below* the point that is struck.

Thus, on the first of May, 1800, a church situated on Mount St. Ursula, a lofty peak in Styria, was struck; and *seven persons* were *killed* by a flash of lightning darting *upwards* from a thunder-storm below.

345. From the observations of Peytier and Hossard among the Pyrenees, it appears, that the upper and lower surfaces of thunder-clouds bear no resemblance to each other, for while the latter are perfectly *level*, the former are *broken* and *uneven*, presenting the appearance of mountains and ridges; whence, during seasons of great heat, lofty peaks and pinnacles of clouds shoot far up into the sky.

LIGHTNING.

346. ORIGIN. When a portion of air is subjected to a very *sudden* and *powerful compression*, a spark is elicited (Art. 551): that electricity produces such a compression can be proved by experiment, and to the energetic *condensation* of the *atmosphere* before the electric fluid, in its rapid progress from point to point, is attributed the vivid flashes that illumine the stormy sky.

347. KINDS. Lightning has been divided by Arago into *three kinds*, principally distinguished by their *form*

What did Peytier and Hossard observe?
What is the cause of lightning?
Into how many kinds has it been divided by Arago?

viz., *zigzag-lightning, sheet-lightning,* and *ball-lightning.*

348. ZIGZAG-LIGHTNING. This kind is so called from the peculiarity of its figure, which is thus explained. As the electricity passes through the atmosphere, the air is supposed, at length, to be so powerfully compressed before it, that a great resistance is presented, and the electric fluid then finds an easier route in some other direction. In this it proceeds, until it once more meets with a like opposition, and is compelled again to change its course; and thus it continues glancing from side to side, until at last it reaches the place it seeks.

Zigzag-lightning appears as a *narrow, jagged line of intensely vivid light,* traversing space with extreme velocity. On account of the unequal conducting power of different portions of the atmosphere, the flash sometimes divides, branching out in several different directions; the lightning is then said to be *forked.* A division into *three* distinct lines is of rare occurrence; but even more have been seen, for Kaemtz beheld, at Halle, in June, 1834, a flash of lightning which threw out numerous branches at the sides; the whole presenting the figure of a spine, with its supporting ribs.

It is said that zigzag-lightnings usually pass between the clouds and the earth, seldom flashing from cloud to cloud.

349. SHEET-LIGHTNING. This kind is the *most common,* and appears during a storm as a *diffuse glow of light,* illuminating the edges of the clouds; and at times breaking out from the central mass. When it occurs, the clouds are said to *open.* The flashes of sheet-lightning often follow each other in rapid succession, for the space of many hours; their intensity is by no means great, and the thunder which attends them is low and distant.

350. BALL-LIGHTNING. Lightning of this class is

What are they?
To what is the peculiar figure of *zigzag-lightning* owing?
What is its appearance?
Describe *sheet-lightning.* Describe *ball-lightning.*

extremely rare, and so singular are its attendant phenomena, that we might well doubt its existence, were not the instances of its occurrence fully authenticated. In a storm that happened at Steeple Aston, Wiltshire, in 1772, the Rev. Messrs. Pitcairne and Wainhouse, while in the vestry of the church, saw suddenly before them, at the distance of a foot, and at about their own height from the floor, a *ball of fire, nearly the size of a man's fist, surrounded by a black smoke*. It burst with an explosion like the discharge of several cannon Pitcairne was dangerously wounded, and his person and clothes showed the usual marks of lightning.

During a thunder-storm that occurred in 1809, at Newcastle on Tyne, the house of David Sutton was struck: the lightning descending the chimney. After the explosion, several persons who were assembled in a room, saw at the door a *globe of fire*, which, after remaining stationary for some time, advanced into the middle of the room, where it burst into fragments, with a report like a rocket.

351. On the fourth of November, 1749, in 42° 48′ N. Lat., 2° W. Long., the crew of the ship Montague beheld, a little before noon, and beneath an unclouded sky, a *globe of bluish fire*, like a *millstone*, rolling rapidly upon the sea. At a short distance from the vessel, it rose perpendicularly from the water, and struck the masts with an explosion louder than the discharge of a hundred cannon. Five sailors were thrown senseless upon the deck, one of whom was severely burned.

In the midst of a storm in Scotland, *two globes* of fire, connected together like chained cannon-shot, were seen by a Mr. Lumsden, passing through the sky revolving one about the other, and striking at last upon the summit of a hill. Philosophers have not yet been enabled to account for lightning of this description; it has, however, been supposed to arise from an *unintermitted* discharge of electricity.

352. HEAT-LIGHTNING. It not unfrequently hap-

Relate instances. How is ball-lightning supposed to arise?
What is heat-lightning?

pens, during the serene evenings of summer, that the horizon is illumined for many hours with successive *flashes of light*, unattended with thunder. This is called *heat-lightning*, and has much perplexed meteorologists. It is affirmed by some, that this illumination is the reflection from the atmosphere of the lightnings of remote storms; the storms themselves being so far distant, that their thunders cannot be heard. Others assert, that during warm, sultry weather, when the air is highly rarefied, its pressure upon the clouds is so much diminished, that the electric fluid can never accumulate upon their surface beyond a certain point, when it escapes in noiseless flashes to the earth.

353. Multiplied observations have proved, that heat-lightning generally originates in the first-mentioned cause; but the instances are by no means rare, when silent flashes of electric light play between the earth and the clouds. These cases occur when the weather is *sultry*, the air being then both *rarefied* and *moist*; two conditions which lessen its non-conducting power; the atmosphere thus becomes an *imperfect conductor* between the clouds and the earth, which are in opposite electrical states, and opposes just sufficient resistance to the passage of the electric fluid as to render it visible.

354. VELOCITY OF LIGHTNING. By a very ingenious piece of apparatus, Prof. Wheatstone, of King's College, London, has been enabled to show that the duration of a flash of lightning is less than the *thousandth part of a second*, and Arago has demonstrated that it does not *exceed the millionth part*.

Now the *duration of a flash*, is the *time it occupies in traversing the space between two clouds, or between a cloud and the earth*; if we then estimate this distance to be equal sometimes to a quarter of a mile, which is a low computation, the velocity of lightning, in such cases, according to Arago, could not be *less* than 250,000 *miles per second*. The electricity developed by the

---

How does it originate?
What is said in regard to the *velocity* of lightning?

electrical machine, has been shown by another beautiful contrivance of Prof. Wheatstone, to possess a speed of 288,000 miles *per second :* the rapidity of lightning is probably not less.

The preceding remarks apply only to lightnings of the first and second class. Ball-lightnings, on the contrary, often move *slowly*, and are visible for many seconds.

355. COLOR. When thunder-clouds are near the earth, the flashes are of a brilliant *white ;* but when the storm is high, and the lightnings play through a rarefied atmosphere, their color, approaches to *violet.* A spark of electricity assumes the same hue, when it is made to pass through the exhausted receiver of an air-pump.

356. EFFECTS OF LIGHTNING. These are precisely similar to those of common electricity in *kind*, though far exceeding them in *degree.* Life is destroyed by the shock, the stoutest trees shivered to pieces, ponderous weights displaced, combustibles inflamed, metals softened and fused, sand vitrified, and iron and steel rendered magnetic. It is needless to multiply instances in proof of these particular points, but a few cases may tend to impress them upon the mind.

357. On the night of the 21st of June, 1723, a tree in the forest of Nemours was struck by lightning. The trunk was split into two fragments, one *seventeen feet long*, the other *twenty-two ;* and though the first required *four* men to lift it, and the second *eight*, yet both of them were hurled to a distance of *seventeen yards.* On the 6th of August, 1809, a flash of lightning struck a house at Swinton, near Manchester. The wall of a building attached to the house was loosened from its foundation a foot below the ground, and raised in a mass to the surface, still maintaining its upright position ; one end of it was moved nine, and the other four feet from its original place. The wall thus moved was *eleven feet high and three feet thick*, and contained 7000 bricks, which, exclusive of the mortar, were estimated to weigh nearly *twenty-six tons.*

What of its *color?* What of its *effects?*

On the 20th of April, 1807, at Great Mouton, in Lancashire, a windmill was struck by lightning; the fluid passed along a large iron chain, the links of which were so *softened*, that by *their own weight* they became *welded* together; and the chain was converted into an inflexible bar of iron.

In Sept. 1845, a house at New Haven, Ct., was struck during a thunder-storm. Several articles of steel were rendered *magnetic*, and a razor, lying in a case near the spot where the lightning entered, was found capable of sustaining a key, weighing *half an ounce*.

358. FULGURITES. When a flash of lightning falls upon sand, its path below the surface is often marked by a *fulgurite*, so called from the Latin word *fulgur*, lightning. *It is a tube composed of sand, vitrified by the action of the lightning.* Fulgurites were first discovered in Silesia, in 1711, and specimens were forwarded to the museum at Dresden, where they are still preserved: they have since been found in great numbers, in Germany, England, and amid the sands of Bahia, in Brazil.

The fulgurite is winding in its form, often throws out lateral spurs or branches, and contracts in size towards the lower extremity, which usually terminates at a spring of water, or in some substance that is a good conductor of electricity.

359. These tubes are generally hollow, the interior surface being coated with a brilliant glass. Their diameters vary from *four-hundredths* of an inch, to *three inches and a half*, and the thickness of their sides from *one-fiftieth of an inch*, to nearly *an inch*.

The branches of the fulgurite, differ in length from *three quarters of an inch* to a *foot*, but the main tube often extends to the depth of many yards. Several of considerable length, which had been taken from the sandy plains of Silesia, were exhibited at London, some

---

Give instances. What are fulgurites?
Where have they been discovered?
What is their form?
State their dimensions.

years ago, by Dr. Fiedler, of Germany. One, discovered at Paderborn, in Westphalia, was *forty feet long*.

360. That these tubes are really produced by lightning, has been proved by actual observation. A number of sailors, being upon the isle of Amrum, in Denmark, saw a flash of lightning fall upon the sand; upon examining the spot, they found a fulgurite: a similar circumstance happened on the borders of Holland. Savart and others have obtained *artificial fulgurites*, by passing powerful electric sparks through powdered glass, and a mixture of sand and salt; tubes were thus formed an inch in length, and the tenth of an inch in thickness, the inner diameter being the twenty-fifth of an inch.

361. VOLCANIC-LIGHTNING. The clouds of smoke, ashes, and vapor, that issue from volcanoes during their eruption, are the scene of terrific lightning and thunder. Pliny the younger, in his letters to Tacitus, mentions the lightning that was seen above Vesuvius, during its eruption, in the year 79, A. D. In that which occurred in 1767, the inhabitants at the foot of the mountain assured Sir William Hamilton, that they were more terrified at the lightning which flashed around them, than by the burning lava, and all the other attendant dangers.

During the eruptions of the same mountain in 1779 and 1794, there appeared, in the midst of the dark volcanic clouds, *globes of fire*, which, bursting like bombshells, darted on every side vivid flashes of zigzag-lightning. In the latter eruption were heard loud and continued *peals of thunder*.

362. The cause of volcanic-lightning is found, in the *rapid condensation* of the vast volumes of heated vapor, which are carried up from the crater of the volcano into the higher and colder regions of the atmosphere.

In like manner, in the midst of water-spouts and

How is it known that they are actually caused by lightning?
In what manner have they been artificially made?
Relate the instances given of volcanic lightning and thunder
How are volcanic lightnings caused?

whirlwinds, an abundant condensation of vapor suddenly occurs, which frequently develops such an amount of electricity, that the lightning here displays itself in all its fearful energy.

363. THUNDER. In consequence of the lightning passing through the atmosphere with an amazing velocity, it leaves a void space behind it, into which the surrounding air instantly rushes, with a ·loud report. This noise is *thunder*.

When the lightning is near the observer, the report is *sharp* and *quick*, but when at a distance, it is *long* and *rolling*.

364. The *rolling* of thunder is frequently occasioned by the reverberations of the sound, from clouds and adjacent mountains; but this is by no means always the case. When the lightning-flash darts to a great distance, such is its velocity, that the thunder may be considered as occurring at *every point of the flash at the same time*. But sound has a progressive motion of 1142 feet per second, and *all the thunder* will not reach the ear at the *same instant*. It will be first heard from the nearest point, in the path of the flash, and later and later from points more remote; and the combined effect will be a *continued peal*.

The *zigzag form* of the flash, and its *division* into several streams, is regarded by Herschel as affording an adequate explanation for all the changes that occur in the sound of the thunder-peal.

365. The *time* that elapses between the lightning and the thunder, enables us to form an estimate of the distance of the former, which is a little more than a *mile* for every *five seconds*. This interval usually varies from *three* to *sixteen* seconds; but cases have occurred, where it has amounted to fifty, and even *seventy-two seconds*.

366. IDENTITY OF LIGHTNING AND ELECTRICITY.

What is the cause of thunder?
How is its rolling occasioned?
How can we estimate the distance of lightning?
How great an interval of time sometimes occurs?

The *resemblance* between *lightning* and *electricity* was noticed by the earlier electricians, Wall, Grey, and Nollet; but their *identity* was first established by Dr. Franklin. The strong points of similarity which convinced him of this fact, were the following.

1st. *Lightning and the electric spark are both zigzag in form.*

2d. *Lightning strikes trees, chimneys, spires, masts of vessels, mountains and elevated points upon the surface of the earth. Electricity is likewise attracted by pointed bodies.*

3d. *Both choose the best conductors.*

4th. *Both ignite combustibles.*

5th. *Both fuse metals.*

6th. *By the action of each, a bad conductor is shivered when struck.*

7th. *Lightning reverses the poles of a magnet, and renders iron magnetic. Electricity does the same.*

8th. *Animal life is destroyed by each.*

9th. *Blindness is produced by both.*

368. Franklin, however, did not stop here. He resolved to test the truth of his reasoning, by drawing lightning from the clouds, and in June, 1752, made the hazardous experiment in the vicinity of Philadelphia.

369. Franklin's Experiment. Having made a kite, by tying the corners of a large silk handkerchief to the ends of two light strips of cedar that crossed each other, and placed upon it a pointed iron wire connected with the string, Franklin went out into a field upon the approach of a thunder-storm, accompanied by his son. When the kite was raised, he attached a key to the lower end of the hempen string; to the key one end of a silk ribbon was now tied, the other being fastened to a post. The kite was thus insulated, and the experimenter, for a considerable time, awaited the result with intense solicitude. A dense cloud passed over, but no indica-

---

By whom was the identity of lightning and electricity first established? What points of similarity did Franklin observe? Relate Franklin's experiment.

tions of electricity appeared upon the string; when, just as Franklin began to despair of success, he beheld the loose fibres of the cord *starting asunder*, and immediately presenting his knuckle to the key he received an *electric spark*. The rain now descending, increased the conducting power of the string, and vivid electric sparks issued from the key in great abundance. By means of the lightning thus obtained, all the common electrical experiments were performed, and the identity of lightning and electricity thus indubitably proved.

370. ROMAS' EXPERIMENT. No sooner was this wonderful discovery made known, than men of science were eager to repeat the experiment.

With a kite eleven feet high and three feet wide, Romas obtained in France the most brilliant and astonishing results. In one instance, when the kite was raised during a storm, such an accumulation of electricity occurred, that streams of electric fire *nine* or *ten feet long*, and *an inch in thickness*, flashed spontaneously from the string, with reports, like those of a pistol. *Thirty* streams of this magnitude burst forth in the space of an *hour*, without counting a multitude of others, *seven feet in length*.

371. RICHMAN'S DEATH. That such experiments are, however, attended with great danger, unless every precaution is strictly observed, is proved by the unfortunate death of Prof. Richman, of St. Petersburg, who was killed by lightning, on the 6th of August, 1753. He had erected, upon the top of his house, an iron rod from which proceeded a chain that entered his study. The whole apparatus was *entirely insulated*. On the day in question while examining the electrometer, as a thunder-storm was approaching, a large globe of blue fire flashed from the conductor to his head, instantly depriving him of life.

Relate Romas' experiment.
What error did Richman commit in the construction of his apparatus?

LIGHTNING-ROD.

372. The invention of the *lightning-rod* for the protection of buildings was the fruit of the brilliant discovery of Franklin. Even before his decisive experiment he had been led to suppose, from the analogies existing between lightning and electricity, that *pointed metallic rods* might possibly disarm the thunder-cloud of its terrific power.

373. In order that the *lightning-rod*, or *conductor*, may afford an effectual protection, regard must be had to the *material* of which it is made, *its size*, and the *mode of erection*.

374. MATERIAL. Wrought iron is usually employed, and forms a good conductor; but copper is preferable, inasmuch as it is less liable to be corroded or fused, and possesses a greater conducting power.

375. SIZE. The rod, if made of iron, should be *three-quarters of an inch in diameter*, and its upper extremity should terminate in *one or more points*. Each of these points (which are usually three in number) ought to be capped with some metal which does not rust, as silver, gold, or platina; for the conducting power of the points, if made of iron, would be weakened by the rust.

376. MODE OF ERECTION. The rod should be *continuous from the top to the bottom;* an entire metallic communication existing throughout its whole length. This law is violated, when the joints of the several parts that form the conductor are imperfect, and the whole is loosely put together. The parts may be screwed one into the other; or the rod may be formed of wires twisted together.

377. The conductor should be fastened to the building by wooden supports, but if masses of metal, as

---

By whom was the *lightning-rod* invented?
To what particulars must attention be directed, that the lightning-rod may afford an effectual protection?
What is said in regard to the material?
To the size? To the mode of erection?

leaden pipes and troughs, are connected with the building, it is best to attach them to the rod by strips of metal; for, unless this is done, lightning may pass from the rod to the metal, and enter the edifice, especially if the rod is in any way defective. By adopting the above precaution, the metallic masses are made a part of the conductor, and if the lightning strikes them, it is conveyed through the rod to the earth.

378. The lower end of the rod should be divided into two or three branches, so bent as to pass away from the building; and it is highly essential that these branches should extend so far below the surface of the ground, as to reach either water or a *permanently moist stratum* of earth. The rod should be surrounded with powdered charcoal, which at once preserves the iron from rust, and facilitates the passage of electricity between the metal and the earth, in consequence of its conducting power. For the same reason, the conductor should be painted with black paint, made of charcoal.

379. EXTENT OF PROTECTION. According to the investigations of M. Charles, *the lightning-rod protects the space around it to a distance equal to twice its height.* Thus, if the conductor extends ten feet above the summit of a house, it affords protection to a circular space forty feet in diameter; the rod being in the centre.

The experience of nearly *one hundred years* has shown that, where the above rules and precautions are observed, an effectual security has been provided against the effects of lightning; so far as human means can avail to disarm the elements.

380. It is an error to suppose that conductors *attract* the lightning towards the building upon which they are erected. They simply *direct* the *course*, and *facilitate* the *passage* of the electricity between the clouds and the earth, when a discharge must inevitably occur, where the building is situated.

How great a space is protected by a lightning-rod?
Have we any proof of the utility of lightning-rods?
Is a building more or less liable to be struck when furnished with a good conductor?

It is indeed highly probable, that a *silent* and *gradual discharge* of a thunder-cloud, is often effected by the points of the rod, and an explosion thus prevented. This is the opinion of Arago, who expressly states, that "*lightning-rods not only render strokes of lightning inoffensive, but considerably diminish the chance of a building being struck at all.*"

381. ELECTRIC FOGS. Fogs are at times highly electrical; a most extraordinary instance is thus related by Mr. Crosse, of Broomfield, whose apparatus has already been described. "Many years since I was sitting in my electrical room, on a dark November day, during a very dense, driving fog and rain, which had prevailed for many hours, sweeping over the earth, impelled by a south-west wind. I had at this time 1,600 feet of wire insulated, which crossing two small valleys, brought the electric fluid into my room. From about 8 o'clock in the morning until four in the afternoon, not the least appearance of electricity was visible at the atmospheric conductor, even by the aid of the most delicate tests. Having given up the trial of further experiments upon it, I took a book and occupied myself with reading, leaving by chance the receiving ball upwards of an inch from the ball in the atmospheric conductor. About four o'clock in the afternoon, while I was still reading, I suddenly heard a very strong explosion between the two balls, and shortly after many more took place, until they became one *uninterrupted stream of explosions*, which died away and recommenced with the opposite electricity in equal violence. The stream of fire was too vivid to look at for any length of time, and the effect was most splendid, and continued without intermission, save that occasioned by the interchange of electricities, for *upwards of five hours*, and then ceased entirely. The least contact with the conductor would have *occasioned instant death*, the stream of fluid far exceeding anything I have ever witnessed, excepting during a thunder-storm."

What instance is given of an electric fog ?

SPONTANEOUS ELECTRICITY.

382. St. Elmo's Fire. When in a darkened room a needle is brought near to the charged conductor of an electrical machine, the point is tipped with a vivid light, caused by the flow of electricity from the conductor to the needle. In the same manner when thunder-clouds approach very near the earth, lightning does not always occur; but the electricity becomes so intense, that it escapes from one to the other by *points* upon the surface of the earth, which then glow with a *brilliant flame.* This phenomenon has received the appellation of St. Elmo's fire. It was known to the ancients by the name of Castor and Pollux, and many instances have been recorded by classic writers. On the night before the battle that Posthumius gained over the Sabines, the Roman javelins emitted a light like torches; and Cæsar relates that during the African war, in the month of February, there suddenly arose, about the second watch of the night, a dreadful storm that threw the Roman army into great confusion, at which time the points of the darts of the fifth legion appeared to be on *fire.*

383. The fire of St. Elmo is often finely displayed upon the masts of vessels. An extraordinary instance, which happened in 1696, is thus related by Count Forbin: "In the night it became extremely dark, and thundered and lightened fearfully. We saw upon different parts of the ship about *thirty* St. Elmo's fires; among the rest was one upon the top of the vane of the mainmast, about eighteen inches long. I ordered one of the sailors to take the vane down, but he had scarcely removed it when the fire again appeared upon the top of the mast, where it remained for a long time, and then gradually vanished." When Lord Napier was on the Mediterranean, in June, 1818, he observed, during a dark and stormy night, a *blaze of pale light* upon the mainmast of his vessel. It appeared near the summit,

---

What is the cause of spontaneous electricity?
What name has been given to this phenomenon?
Relate the several instances.

and extended about three feet downward, flitting and creeping around the surface of the mast. The heads of the other two masts presented a similar appearance. At the end of half an hour, the flames were no longer visible.

384. This phenomenon frequently occurs on the *summits of mountains*, when thunder clouds pass near them. Saussure observed it upon the Alps, in 1767. On extending his arm, he experienced slight electric shocks, accompanied by a whistling sound, and obtained *distinct sparks* from the gold button of a hat belonging to one of his party. It is often noticed at Edinburg castle, which stands upon a high rock, 250 feet above the surrounding country. Upon the approach of a storm, the bayonets of the soldiers mounting guard are frequently seen capped with flame, and an iron ramrod, placed upright upon the walls, presents a like appearance.

A singular instance of spontaneous electricity took place at Algiers, on the 8th of May, 1831. During the evening of this day, as some French officers were walking with their heads uncovered, each was surprised at seeing the hairs upon the heads of his companions *erect*, and *tipped with flame.* Upon raising their hands, they perceived a similar light flitting upon the ends of their fingers.

A remarkable case of this kind was observed by Pres. Totten, of Trinity College, at Hartford, Ct., in the month of Dec. 1839. As this gentleman was walking one evening in the midst of a heavy snow-storm, protected by an umbrella, his attention was arrested by momentary *flashes of light*, which at intervals illumined his path. The source of the light was detected upon meeting another person, the point of whose umbrella was seen covered with *flame*, which was constantly escaping in *flashes*. The light first noticed by Pres. Totten, proceeded from his own umbrella.

385. ELECTRIC RAIN, HAIL, AND SNOW. Numerous

When does it occur on the summits of mountains?
State the cases in Art. 384.

and well attested instances have occurred, in which rain, hail, and snow, have displayed flashes of electric light, but we will confine ourselves to a few. On the 22d of Sept. 1773, in a thunder-storm which fell upon Skara, in Sweden, the *drops of rain* were seen to *strike fire* and *sparkle* as they touched the ground.

On the 28th of Oct. 1772, as the Abbe Bertholon was traveling between Brignai and Lyons, in the midst of a heavy storm, he was surprised at seeing the *rain-drops* and *hail-stones* emitting *jets of light*, as they fell upon the metallic parts of his horse's trappings.

It is also recorded, that the miners of Freyburg, on the 25th of January, 1822, beheld the *sleet* which fell during a storm *flash with light* as it struck the earth.

386. It is not difficult to explain these phenomena; we have only to suppose, that the electric intensity of the atmosphere and the earth is at these times very great, and that the electricity of the falling bodies is the opposite in *kind* to that of the ground and of the objects upon it. At the moment of contact the two kinds of electricities combine, their union (as is always the case when their intensity is great) being indicated by a sudden flash.

387. ELECTRIC ACTION UPON TELEGRAPHIC WIRES. It is not unusual for the electricity of the atmosphere to exert an extraordinary influence upon the wires of the electrical telegraph. According to Prof. Henry, this influence may arise, as follows, in several different ways.

388. First. *The wire may be struck by a direct discharge of lightning from the clouds.* An instance of this kind occurred on the 20th of May, 1846, when the lightning struck the wire of the telegraph, at the place where it crosses the Hackensack river. From the point

---

Relate the instances given of electric rain, hail and snow.
How are these facts explained?
What are Prof. Henry's views respecting the influence of atmospheric electricity upon telegraphic wire?
What is the first mode of action?
Give the instance.

where the discharge took place, the fluid passed along the wire each way for a distance of several miles, striking off at irregular intervals down the supporting poles. Wherever a pole was struck, a number of sharp explosions were successively heard, like the rapid reports of several rifles.

389. Secondly. *The state of the wire may be disturbed by the conduction of a current of electricity from one portion of space to another, without the presence of a thunder-cloud;* and this will happen in the case of a long line, when the electric condition of the atmosphere which surrounds the wire at one place is *different* from that at another.

390. This difference in the electric condition of the atmosphere may result from a *difference in elevation.* (Art. 322.) A wire, raised by means of a kite, gives sparks of positive electricity, in a perfectly clear day ; hence, if the telegraphic wires pass over a high mountain-ridge, a current of electricity will be continually flowing, during serene weather, from the more elevated to the lower parts of the wire.

A current may also arise in a long, level line, if a *fog* exists at one end, while the sky is clear at the other ; or if a *storm of rain or snow* occurs at some portion of the line, while the remainder is free from its presence.

Currents of electricity have been produced by some of these causes, of sufficient power to set in motion the working machine of the telegraph. In one case it began to operate *spontaneously*, without the aid of the battery, when a snow-storm prevailed at one end of the line, and clear weather at the other.

391. Thirdly. *The inductive action of a thunder-cloud may also change the natural electric condition of the wire.* If, for example, a cloud *positively electrified*, is moving across the direction of the wire, it will con-

---

What the second ?
When will this disturbance happen in the case of a long line ?
How may this difference arise ?
What is stated respecting the energy of the currents thus produced ?
What is the third mode of influence ?

tinue, as it gradually approaches the line, to drive to the remote extremities of the wire, more and more of the positive electricity residing in it, and thus occasion a *current*. As the cloud gradually recedes, the repulsion it exerts is diminished, and a *current* then arises in the *opposite direction*.

392. Fourthly. *Every flash of lightning which occurs within many miles of the line, produces powerful electrical currents in the telegraphic wires.*

To this influence Prof. Henry attributes the phenomena witnessed by himself, on the 19th of June, 1846, in the telegraph office, at Philadelphia, and which he thus describes. "In the midst of the hurry of the transmission of the congressional intelligence from Washington to Philadelphia, and thence to New York, the apparatus began to work irregularly. The operator at each end of the line announced at the same time a storm at Washington, and another at Jersey City. The portion of the telegraphic wire which entered the building, and was connected with one pole of the galvanic battery, happened to pass within the distance of less than an inch of the wire, which served to form the connection of the other pole with the earth. *Across this space*, at *intervals* of every *few minutes*, a series of sparks in rapid succession was observed to pass; and when one of the storms arrived so near Philadelphia that the lightning could be seen, each series of sparks was found to be *simultaneous* with a flash in the heavens. Now we cannot suppose, for a moment, that the wire was actually *struck* at the time each flash took place, and indeed it was observed that the sparks were produced, when the *cloud* and *flash* were at the distance of *several miles to the east of the line of the wire*. The inevitable conclusion is, that all the exhibition of electrical phenomena witnessed during the afternoon, was purely the effect of induction, or the mere disturbance of the natural electricity of the wire at a distance, without any transfer of the fluid from the cloud to the apparatus.

What the fourth?
Describe the phenomena witnessed by Professor Henry.

" The discharge between the two portions of the wire continued for more than *an hour*, when the effect became so powerful, that the superintendent, alarmed for the safety of the building, connected the long wire with the city gas pipes, and thus transmitted the current silently to the ground."

393. By a simple apparatus, Professor Henry rendered manifest the inductive action of the lightning-flash. The arrangement consisted of a copper wire, fastened at one end to a building, and extended to another, 400 feet distant. Here it entered the Professor's study, and thence passed through a cellar window into an adjoining well. With every flash of lightning that occurred within *a circle of twenty miles*, needles were magnetized in the study by the induced current of electricity developed in the wire. (C. 1033.)

What apparatus was constructed by this gentleman for exhibiting the inductive action of the lightning?

What effect was produced?

PART V.

OPTICAL PHENOMENA.

CHAPTER I.

OF THE COLOR OF THE ATMOSPHERE AND CLOUDS.

394. COLOR OF THE ATMOSPHERE. This is caused by the decomposition of the solar light. It is well known, from the experiment of the prism (C. 788), that the white light of the sun consists of seven colors; and, that of all these, the *violet and blue* rays have the *least power* to *overcome* any *resistance* they meet with; and consequently *deviate most* from their original course in passing through the prism.

395. The action of the atmosphere upon the sunbeams in their passage to the earth, is precisely similar to that of a prism.

After entering the atmosphere they are constantly passing in their onward progress, from rarer into denser media, and are therefore decomposed. A portion of the blue rays, unable to overcome the resistance of the air are scattered throughout its extent; and being reflected from its particles, tinge the sky with an azure hue; for it is to be remarked, that *a body appears of the same color* as the light *it reflects* and by which it is *seen*.

396. CYANOMETER. In order to determine the intensity of the blue, a *cyanometer* is employed, an instru-

---

What is the subject of part fifth?
Of what does chapter first treat?
What is the cause of the color of the atmosphere?
For what purpose is the cyanometer employed?

ment which derives its name from the Greek words, *kuanos*, azure, and *metron*, measure. That of Saussure is made in the following manner:

A circular card is divided into *fifty-one* parts, and each is painted of a different shade of blue increasing from the palest tint, formed by a union of blue and white, to the deepest produced by a mixture of blue and black. The colored card being held in the hand, the observer marks the particular tint corresponding to the color of the sky, and its number, counting from the palest shade, denotes the intensity of the azure.

397. EFFECT OF LATITUDE. *The brilliancy of the sky decreases with the latitude.* Humboldt discovered that at corresponding heights above the horizon, the blue in 19° N. Lat. was two shades below that in 16° N. Lat. The intensity also at Cumana, 10° N. Lat., is *twenty-four*, while the average tint for Europe is only *fourteen*.

398. This diminution in brilliancy is caused by the less perfect absorption of the atmospheric humidity in the temperate and arctic regions, than in the equatorial climes—a circumstance arising from their comparative low mean temperature, and consequent decrease in the capacity of the air for moisture.

399. In the *same place*, the color *increases* in *brightness* from the *horizon* to the *zenith*—the point in the heavens directly over-head. Baron Humboldt found in 16° N. Lat., that his cyanometer indicated the 3d shade at the horizon, but at an altitude of 60°, or two-thirds of the distance to the zenith, the 22d tint. The blue of the sky is palest at the horizon, in consequence of being mixed with and diluted by the thin vapors of the air, which settle down towards the earth.

Within the torrid zone, the sky undimmed by vapors glows with the purest azure; and, under like circumstances, in regions beyond the tropics, the same bright

---

Describe this instrument.
How is the brilliancy of the sky affected by latitude?
State the result of observations. How is this difference accounted for?
What is the law in respect to altitude at the same place?
Give the observations. What is said of the torrid zone?

skies are seen. Over Italy, California and the Canary Isles, hangs a canopy of the deepest blue, and even on the western coast of Spitzbergen, the rich azure of the heavens has equaled at times the splendid hue of the tropic skies.

400. EFFECT OF ALTITUDE. In ascending from the plains to the mountains, the vapors are left below, the purity of the atmosphere increases, and the *pale tint* of the sky *changes to a vivid blue*. This fact, long known to the chamois hunters of Switzerland, was verified by the observations of Saussure upon the Alps, and those of Humboldt on the Cordilleras.

401. Capt. Mundy thus speaks of the color and pureness of the air at Simla, which is the most northern European settlement in India, and possesses an altitude of 7,800 feet. " To the north of Simla, the mountains rise gradually one above another, until the panorama is majestically terminated by the snowy crescent of the great Himalaya belt, fading, on either hand, into indistinct distance. In fine weather, these stupendous icy peaks cut the dark blue sky with such sharp distinctness of outline, that their real distance of *sixty* or *seventy* miles is, to the eye of the gazer, diminished to *one-tenth part*."

402. Brantz Mayer, in his interesting work upon Mexico, thus alludes to the same facts.

" The moonlight of Mexico is marvelously beautiful. That city is 7,500 feet above the level of the sea. The light comes pure and pellucid from heaven. You seem able to touch the stars, so brilliantly near do they stand out, relieved against the back-ground of an intensely blue sky. Strolling on such a night in Mexico, I saw the sharp lines of tower and temple come boldly out with shape and even color almost as bright as, yet softer than at noon-day."

403. At Mussoori, a village situated upon the first

What is said of Italy, California, the Canary Isles and Spitzbergen?
What is the law in regard to altitude at different stations?
What is said of the blueness of the sky and the purity of the atmosphere on the Alps and the Cordilleras, on the Himalayas and in Mexico?

range of the Himalayas, 7,500 feet above the sea, so remarkably clear is the air in the month of November, according to Lieut. Bacon, that the white houses at Moozaffirnuggur, a distance of *eighty-two miles*, have been distinctly seen with the aid of a spy-glass.

404. When, however, very lofty elevations are attained, the heavens assume a *blackish hue;* for a great portion of the atmosphere is then beneath the observer, and but little blue light is reflected from the comparatively small number of particles composing the attenuated air above. The celestial orbs there shine with a singular brilliancy, since their light reaches the eye before its lustre has been dimmed, in consequence of passing through the dense strata of the atmosphere near the surface of the earth.

405. Captain Hodgson remarked, near the sources of the Ganges, that the tint of the sky was a *dark blue*, approaching to *blackness;* and that the stars in their rising emerged with a sudden flash from behind the snowy peaks of the Himalayas. At Zinchin, *sixteen thousand* feet above the sea level, the heavens appeared of a *dark black* color, the sun shining without the least haze. At night, that part of the horizon, where the moon was expected to rise, could scarcely be distinguished by the irradiation of her beams before the orb touched it, and the stars and planets shone with a dazzling light.

406. COLORS OF CLOUDS. These are attributed to the power which the atmosphere possesses of *absorbing light* (C. 802), in common with other transparent media. When a sun-beam falls upon the ocean, the more refrangible rays are successively absorbed as the light continues to pierce the translucent water; until, at last, far beneath the surface, nothing but *red* light is perceived, according to the statements of divers, and of those who

---

What is the color of the sky when very lofty heights are attained?
Relate the remarks of Capt. Hodgson.
Explain the cause of the colors of clouds, and the manner in which they arise.

have been engaged in submarine researches. The action of the atmosphere is precisely the same.

In the morning and evening, the sun-light traverses the densest portions of the air, and passes through a onger track than at any other time. So much thicker and more dense is the stratum of air upon the horizon than the stratum over-head, that the sun-light is diminished 1300 times in traversing the former; and of 10,000 rays falling towards the surface of the earth, 8,123 arrive at a given point if they pass perpendicularly through the air, but *only five* if they come through a horizontal stratum. From these causes, the more refrangible rays, and especially the violet and blue, are unable to struggle through and are absorbed; while the rest *emerge*, and being *reflected* from the light masses of vapor floating in the sky, clothe them with their own bright hues.

407. The three most powerful rays of the solar spectrum are *red*, *orange*, and *yellow*; and these colors are the common tints assumed by clouds. At times, however, they glow with the richest variety of hues, particularly beneath the tropic skies. In those regions, *green*, *violet* and *purple* clouds are not of unfrequent occurrence.

Bishop Heber, on his passage to India, beheld, one evening at sunset, when near the equator, large tracts of cloud of a *pale, translucent green*, surpassing in beauty every effect of paint, glass, or gem.

The sunsets of California are among the most beautiful in the world, and the clouds that rise from the Pacific are bathed in exquisite tints of green, purple, and violet.

408. Clouds, possessing these singular colors, are rarely seen in the higher latitudes; they are however not entirely unknown. Violet clouds have been witnessed at Avignon, in France, and also in a most gorgeous sunset that occurred at Hartford, Ct., on the 3d of July, 1844, and which presented the following phenomena.

409. The day had been showery, but towards its

What is said as to the diminution of light?
In what regions are the richest tints beheld? Give instances.
Where are violet and green clouds most frequently seen?

close, the dense canopy of clouds was broken up, and the eastern sky filled with light and floating masses of vapor. Soon after sunset, the stratum of clouds which rested upon the western horizon, rose throughout its whole length, revealing between the mountains and its lower edge a belt of sky of the purest azure. Above this, the whole field of vapor was gleaming with a rich *amber* light, which, as it streamed through rarer or denser portions of the mass, presented every phase of brilliancy and depth ; at the same time displaying the curiously wrought structure of the airy fabric. When the rays of the sun fell upon the fragments of vapor floating in the eastern quarter of the heavens, their jutting heads and broken edges gleamed with a flame-like hue ; while, between the masses, the sky appeared of the *deepest indigo*. As the evening advanced, portions of the western stratum assumed the tints of *lead, lake, pink, green, purple, violet, orange* and *crimson*. About eight o'clock, the vapor in the south-west presented a singularly beautiful appearance ; the heavens seemed as if covered with a delicate *lace-work woven of prismatic rays*, and this phenomenon was succeeded by *green, purple,* and *violet* clouds in the west. The last hue of this brilliant pageant was an *intensely vivid crimson*, which was gradually lost in the shades of night.

410. At the same city, in May, 1845, a cloud of uncommon beauty was seen by the writer resembling *marble paper*, the intermingling colors consisting of *bronze, orange,* a *vivid grass green,* and a *golden yellow*.

411. *Green* clouds occur, when the vapor is illuminated at the same time by the deep blue light, reflected from a distant quarter of the sky, and the yellow rays of the sun ; green being produced by the union of *blue* and *yellow*. In the same manner, *purple* and *violet* clouds appear, when they glow at once with the *red rays* of the sun and the *azure tint* reflected from the heavens ; since purple and violet arise from a mixture of red and blue in due proportions.

Describe the sunset at Hartford.
Explain the origin of green and violet clouds.

CHAPTER II.

OF THE RAINBOW.

412. *The rainbow is that beautifully colored arch which, at times, is seen during a shower and in the region of the sky opposite to that where the sun is shining.*

413. When perfect, the rainbow consists of *two arches*, the *inner*, called the *primary bow*, and the *outer* the *secondary*, each composed of seven colored bows, formed of the prismatic hues, viz., *violet, indigo, blue, green, yellow, orange*, and *red*. In the primary bow the *red* ring occupies the *highest* place, the *orange* comes next, and so on, the *violet* assuming the *lowest* position: but in the secondary the order of colors is *reversed*.

414. The cause of the rainbow, with the exception of its colors, was first unfolded by Descartes; but the discovery by Newton, of the different refrangibility of the sun's rays, enabled this great philosopher to explain, with the utmost completeness, all the laws of this brilliant phenomenon.

415. In order to understand the theory of the rainbow, we must have recourse to diagrams.

Imagine, in the first place, that P Q L, figure 20., is a section of a globe of water, and that S P is a ray of light, which, passing through the hole of a window-shutter in a darkened room, falls upon the surface of the globe at P. Here a portion of the light is *reflected*, and the rest, entering the globe is *refracted* (C. 702) and decomposed into the seven primary colors; *one* of which, the *red ray*, we will

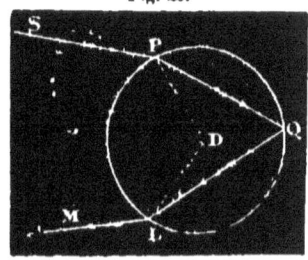

Fig. 20.

SECTION OF A GLOBE OF WATER.

What is the subject of chapter second?
Define the rainbow.
Of what does it consist when perfect?
Who first explained the cause of the rainbow?
Trace the course of a ray of light in figure 20.

now alone consider. This ray, traversing the water, strikes the interior surface of the globe at Q; where a part of its light is lost by *transmission*, and escapes into the air, while the remainder is *reflected* to the point L; here the light is once more subdivided, one portion being *refracted* to the eye situated in the direction L M, and the other *reflected* into the globe.

That the results are such as have been described may be ascertained by placing the eye at the point Q of the globe, and observing likewise the course of the ray through the air and water.

416. These successive transmissions and reflections are unlimited in number, and, since light is lost at each impact, the intensity of the ray, after a few reflections, will become so much diminished, that, upon its emergence from the globe, it ceases to make any impression upon the eye.

417. In order to apply these remarks to the subject before us, we have only to suppose, that figure 20. is the section of a *rain-drop*, instead of the section of a globe of water; for all the changes which light undergoes in one case, will take place in the other.

Thus the ray of light S P, falling upon the drop at P, is *refracted* towards the perpendicular P D, *reflected* at Q, and *refracted from* the perpendicular L D as it passes into the air at L, to meet the eye in the direction L M: all these results occurring, in accordance with well-known optical laws. (C. 706, 709.)

418. In the case supposed, the ray S P suffers *two refractions* and one *reflection*, and if it strikes the drop more or less obliquely, different quantities of light will be brought to the eye at M. Now it is only when the *greatest amount* is conveyed to the eye that the light is sufficiently intense to produce any impression upon the sight, and this is found to occur, in respect to the *red*

Apply the illustration to the subject.

In the case supposed, how many refractions and reflections does the ray undergo?

Under what circumstances only does the emergent ray make any impression upon the sight?

ray, when the angle S B M, figure 21., made by prolonging the lines of the incident and emergent rays S P and M Q till they meet at B, and called *the angle of deviation,* is equal to 42° 2'. *

This angle of greatest intensity varies however for each prismatic color, being 40° 17' in the case of the *violet,* and increasing, for each hue, from the violet to the red.

419. PRIMARY Bow. If we consider P Q L D, figure 21., to be a section of a rain-drop; of all the rays that fall upon it from any *one point* in the sun, some, as S P, will so strike it, as to meet the eye of the observer (supposed to be at M) with the *greatest*

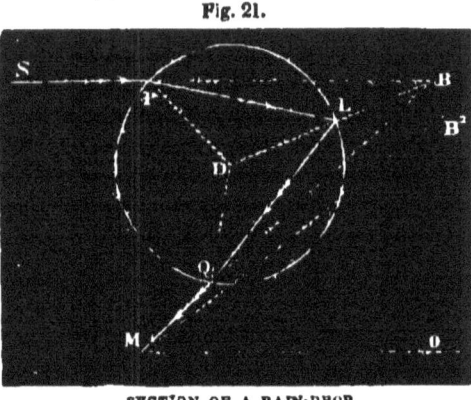

Fig. 21.

SECTION OF A RAIN-DROP.
One Reflection—two Refractions.

---

* An *angle* is the *opening* between two straight lines that meet each other.

Thus the opening between the straight lines, A B and C B, which meet at B, is called the angle B, or the angle A B C; the letter at the point of meeting always being placed in the middle. The size of an angle is computed as follows. The circumference of any circle being divided into 360 equal parts, each part is called a *degree;* a degree being divided into 60 equal parts, each part is called a *minute;* and a minute being divided into 60 equal parts, each part is called a *second.* If now we take B as the centre of a circle,

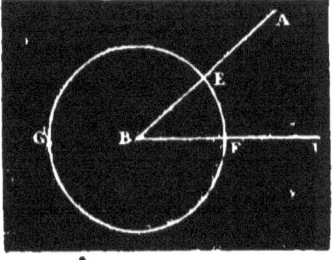

and describe the circumference, G E F, cutting the two lines, A B and C B, in any two points, as E and F, the number of degrees, minutes and seconds contained in the part of the circumference, E F, included between the two lines, A B and C B, gives the value of the angle, A B C. For example, if the length of the circumference, G E F, was so great that it measured 360 inches, and the part, E F, contained 40 *inches* and *nine-sixtieths* of an inch, A B C would be angle of *forty degrees and nine minutes* (40° 9'). Degrees are designated by the following character, °; minutes thus, '; and seconds thus, ".

What is the angle of deviation for the red ray, when the greatest amount of light comes to the eye?

What is the angle of greatest intensity for the violet?

Explain figure 21.

possible brilliancy: making the angle S B M equal to 42° 2', in the case of red light. If the line M O is now drawn parallel to S B, it may be regarded as a ray of the sun *passing through the eye of the spectator;* and since all the rays of the sun are parallel to each other at the earth, the angle B M O will also be equal to 42° 2'.

420. The observer then being placed with his *back* to the sun, and his *eye* at M, will receive the impression of red light from the drop P L Q D, in the direction B M; and not only from this drop, but also from every other drop, whose angular distance from the line M O is, at that moment, the same.

It is therefore evident, if we suppose the line M B to turn about M O, like the legs of a pair of compasses, that *all the points at which red light is seen lie in the circumference of a circle whose centre is O; and that around this centre an arch of red light will appear in the heavens.*

421. The breadth of this arch will be equal to the apparent diameter of the sun, or about 32'; for what has been said in regard to rays proceeding from any *one point* in the sun, viz., that some of them will reach the eye under the angle of greatest brilliancy, is equally true of those which emanate from *every point* of his disk.

422. The explanation of the origin of the red arch is equally applicable to the rest of the colored arches. The latter will be found, however, *below* the former; for, since their angles of greatest brilliancy are each less than that of the red, they must consist of portions of *smaller, concentric circles.*

Thus, the violet arch can only be seen from drops *below* and *within* B, when the light that meets the eye coming in the direction B^2 M, makes the angle B^2 M O equal to 40° 17'. Between the violet and red arches the other colored bows will be seen arranged in the order of the spectrum; the whole forming, by their union, the *primary bow.*

Explain the manner in which the red arch is caused. What is its breadth?
Apply the same mode of reasoning to the other colored arches.
How is the primary bow formed?

OPTICAL PHENOMENA.

423. SECONDARY BOW. The secondary bow is formed when the sun's rays, entering the bottom of the drop, suffer *two reflections* from the interior surface, and emerging at the top, reach the eye of the spectator after *two refractions.*

The course of the ray is seen in figure 22., where S E A is the *incident* ray, B and C the *two points* of *reflection,* and D E H the *emergent* ray, supposed to meet the eye of the observer at H.

Fig. 22.

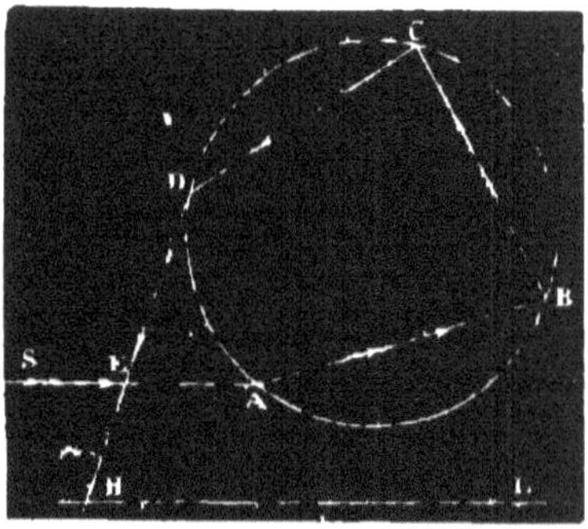

SECTION OF A RAIN-DROP.
Two Reflections—two Refractions.

424. So much light is lost by these successive changes in direction, that only at certain inclinations a sufficient quantity reaches the eye, from each of the prismatic colors, to produce the secondary bow. Its tints, after all, are faint compared with those of the primary.

425. The *violet* light can only be seen when the *angle of deviation* S E H is 54° 9', and the *red* when it is 50° 59'. Suppose, as in the case of the primary bow, that H L is the direction of a ray from the sun passing through the eye of the observer, and making with H E the angle L H E equal to the angle of deviation. If then, the line H E revolves about H L, the spectator, with his back to the sun and his eye at H, will behold in the heavens, between the limits of 54° 9' and 50° 59' a *prismatic bow* consisting of similar portions of *seven concentric circles ;* the *violet arch* assuming the *highest position* and the *red* the *lowest.* The other colors occupy intermediate places; the greater their *refrangibility* the greater their *elevation.*

Under what circumstances does the secondary bow occur?
Trace the course of the ray in figure 22.
What is said of the brilliancy of the secondary bow?
What must be the size of the angle of deviation that the violet light can be seen? What the size that the red ray may be visible?
How is the secondary bow formed?

426. The subject is further illustrated by the following figure, where the four parallel lines represent rays of the sun falling upon four drops of water, and O P. the direction of another ray imagined to pass through the eye of the spectator, R O and V O are the *red* and *violet* rays of the *primary bow;* R' O and V' O the red and violet rays of the *secondary;* and the positions of the *red* and *violet arches* of the two bows are indicated

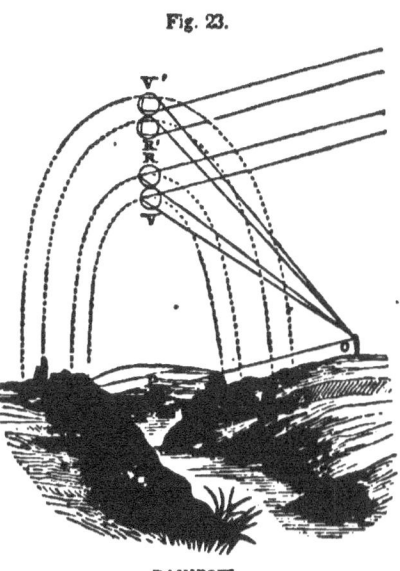

Fig. 23.

RAINBOW.

by the *dotted lines.* The other colored arches are found between the red and violet, following the order of colors in the prismatic spectrum. P is the centre of the rainbow.

427. In the explanation just given, we have reasoned as if the rain-drops were stationary, which of course is not the case; but this supposition leads to no error, inasmuch as the air is filled with rain-drops during the prevalence of a shower, and before one set of drops, by sinking too low, ceases to present to the eye the colors of the bow, another set has descended, taken their place, and is performing their office. While the *observer is stationary* the *rainbow is fixed in position*, but the *drops* that give rise to its glowing tints are *continually changing*.

428. BREADTH OF THE Bows. The angular distance from the middle of the red arch to the middle of the violet in the inner bow, is the difference between 42°

Illustrate farther from figure 23.
Why is the bow stationary although the drops are in motion?

2' and 40° 17' or 1° 45'; a quantity nearly equal to three and a half times the apparent breadth of the sun. This space is occupied by the remaining five colored arches, and, as each is 32' in width, (Art. 421,) they necessarily *overlap* one another, and cause, by their mutual blending, an indistinctness in the boundary of the several hues. The two half-breadths of the red and violet arches added to 1° 45' give the whole width of the bow, which is equal to 2° 17', or *about four and a half times the apparent diameter of the sun.*

429. The breadth of the exterior bow, from the middle of the red to that of the violet, is found in like manner to be 3° 10'—the difference between 54° 9' and 50° 59'. To this quantity 32' must be added to obtain the entire breadth.

The interval between the bows, computing from the red of the primary to that of the secondary, is 8° 57'. All these results, deduced *theoretically, precisely agree* with those obtained by *actual measurement.*

430. POSITION AND SIZE OF THE RAINBOW. Since the centre of the rainbow is in the direction of the line imagined to be drawn from the sun through the eye of the spectator, its *position* will evidently vary with that of the spectator, and its *size* with the altitude of the sun. If this luminary is 42° 2' above the horizon, the top of the inner bow will be *just visible;* but if upon the horizon, the bow will be a *semicircle*, having an elevation of 42° 2'. If the observer, in the latter case, were upon the summit of a mountain, the arch would be somewhat *greater* than a *semicircle;* since the line of direction from the sun through his eye, would strike the sky opposite, at a point *above* the horizon.

Should a person happen to be upon a mountain, when the sun is high in the heavens, and a shower at the same time occur in the vale below, he will sometimes perceive a rainbow forming a *complete circle.*

State what is said in regard to the breadth of the bows.
What in respect to the position and size of the rainbow.
When are entire circles beheld?

Such are said by Ulloa to be frequently seen on the mountains of Peru above Quito.

The foaming waters of cataracts are often spanned by richly tinted bows, caused by the rising spray. They are regularly seen at the falls of Schaffhausen, on the Rhine, and at the cataract of Niagara. At Terni, in Italy, where the river Velino rushes over a precipice 200 feet high, a bow of rare beauty is beheld. It appears, to a spectator below, arching the falls with its glowing tints, while two other bows are reflected on the *right* and *left*.

431. RAINBOWS IN THE NORTH. Rainbows are sometimes seen at *mid-day*. On the 13th of Dec. 1847, at one o'clock, P. M., Prof. Olmstead beheld at Yale College an entire bow in the north. During the same week, the writer observed at Hartford a similar bow at nearly the same hour of the day. Such a phenomenon can never arise, in the case of the primary bow, unless the sun's altitude at the time is considerably less than 42°, which only happens in the winter.

432. EXTRAORDINARY Bows. When the light of the sun is reflected from the surface of tranquil water, rainbows of singular form are at times observed. On the 6th of August, 1698, Dr. Halley beheld, while walking on the walls of Chester, by the river Dee, a rainbow of the form represented in figure 24., where A B C is the primary bow, D E F the secondary, and A H G C the extraordinary bow, cutting the secondary at H and G. Its colors were arranged like those of the primary.

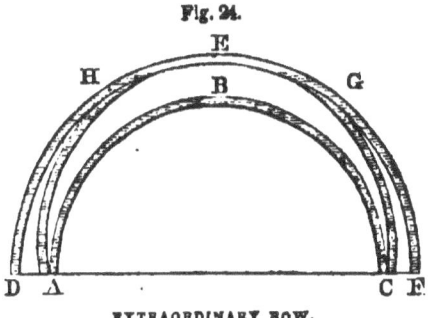

Fig. 24.

EXTRAORDINARY BOW.

Give the instances of rainbows over cataracts.
When can rainbows appear in the north?
Explain from figure 24. the extraordinary bow seen by Halley.

433. The sun was shining clearly upon the calm surface of the river, and Dr. Halley discovered that the extraordinary bow was nothing more than the rest of the circle of which the primary was a part, bent upwards by *reflection* from the water.

A similar rainbow, formed by reflection from the river Eure, was beheld at Chartres, in 1665; when a faint arch was seen crossing the primary at its summit.

434. SUPERNUMERARY Bows. Arches of prismatic colors are sometimes seen, both *within the primary*, and *without the secondary* bows, to which the name of supernumerary or supplementary bows is given.

435. On the 5th of July, 1828, Dr. Brewster saw *three* supernumerary bows *within* the primary, each composed of *green* and *red* arches. *Outside* of the secondary a red arch was clearly seen, and beyond this a faint green one.

At Montreal, in September, 1823, three supplementary bows were noticed by Prof. Twining, within the primary; exhibiting however, only a single color, which was violet or rather a dull red.

At Hartford, Ct., on the 5th of August, 1847, at sunset, two supernumerary bows were seen by the writer, within the primary, *extending throughout the whole semicircle*. The first, in contact with the primary, consisted of green and red arches, and the second of a single band of pale red light.

The most remarkable phenomenon of this kind, was that observed by the Rev. Mr. Fisher, in Dumfrieshire, and related by Dr. Brewster, at a meeting of the British Association, in 1840. In this case the primary was attended by *five* supernumerary bows, and the secondary by *three*. Kaemtz remarks, that it is not easy to account for these supplementary bows in a satisfactory manner; but according to Young, Arago, and others they arise from the action of the rays of light upon each other: the explanation however, is too abstruse to be here introduced.

436. LUNAR Bows. Rainbows are sometimes pro-

Relate the cases given of supernumerary bows.

duced by the light of the moon; their occurrence, however, is extremely rare, and their tints so very faint as to be scarcely perceptible. One of the most brilliant ever beheld, was seen by Mr. Tunstall, at Gretna Bridge, in Yorkshire, on the night of the 18th of October, 1782. It became visible about nine o'clock, and continued, with varying degrees of brightness, till past two. At first it appeared as a distinct bow *without colors*, but afterwards the tints were very conspicuous and vivid, preserving the same order as in the solar bow, though paler; the *red, violet,* and *green* being the *brightest*. At twelve o'clock it attained its greatest splendor. This phenomenon occurred three days before the moon was full; during its continuance, the wind was very high, and a drizzling rain fell for most of the time.

Another bow was seen by the same observer, on the 27th of February, in the same year. The colors were tolerably distinct, but the orange appeared to predominate. A lunar bow with colors, was also noticed near Chesterfield, about Christmas, in the year 1710, and is described by Thoresby in the Philosophical Transactions.

CHAPTER III.

OF MIRAGE.

437. When a ray of light, proceeding from any object, passes obliquely out of one medium into another of a different density, it is *refracted*, or bent from its course, (C. 704,) and when it reaches the eye, the *object is seen in the direction of the last refracted ray.*

Relate the several instances of lunar bows.
What is the subject of chapter third?
In what direction is an object seen, when the rays that come from it to the eye first pass through media of different densities?

438. Thus, if E represents the earth, and 1-2, 2-3, 3-4, different strata of the atmosphere, decreasing in density from 1 to 4, a ray of light proceeding from the star S, and meeting the exterior stratum of the atmosphere at 4, will be successively

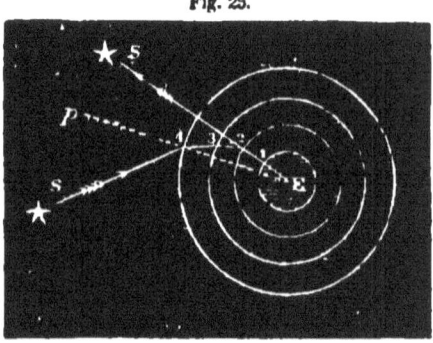

ATMOSPHERIC REFRACTION.

refracted in the directions 4-3, 3-2, and 2-1; so that a spectator at 1 will not see the star S in its real position, but in the direction of 1-2 S'. For this reason all celestial objects, (unless in the zenith, where there is no refraction,) appear *above* their true position. (C. 703.) Thus, the sun and moon, for instance, at their apparent rising and setting are *actually below* the horizon.

439. The variations in the density of the atmosphere near the earth, produced by local changes in temperature, occasion a similar displacement of terrestrial objects; this is *ordinarily* seen in the slight elevation of coasts and ships, when viewed across the sea, and is then called *looming;* but to the more *extraordinary* phenomenon of this nature, the name of *mirage* has been given. When this phenomenon occurs, images of ships erect and inverted are seen in the air, delightful visions of tranquil lakes and verdant fields delude the fainting traveler of the desert, and sometimes, as in the case of Reggio, a noble city with all its splendid panorama of towers and arches, stately palaces and terraced heights, appears like a fairy scene upon the slumbering waters of the sea.

Explain the effect of atmospheric refraction from figure 25.
What is *looming?*
What is *mirage?*

440. INSTANCES.. On the first of August, 1798, Dr. Vince observed, at Ramsgate, a vessel in the distance, the topmast only being visible above the horizon, as at A, fig. 26. Two complete images of the vessel were seen at the same time in the air, the one at C *erect*, and the other below at B *inverted:* between them a distinct image of the sea appeared at D E. The two images were still visible when the real ship had passed *entirely out of sight*.

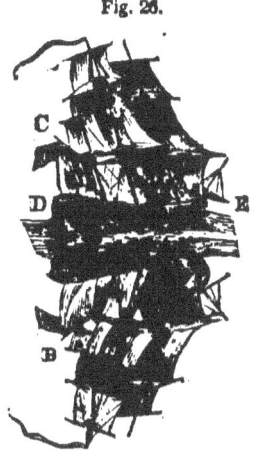

Fig. 26.

441. Similar phenomena were noticed by Capt. Scoresby in 1820, while navigating the arctic seas. In one instance he beheld from the mast-head eighteen sail of ships, at the distance of twelve miles; one appeared *taller* than its actual height, another *shorter;* and above several of the rest, inverted images were seen. In 1822, he recognized his father's ship, the Fame, by an inverted image of the vessel in the air, though it was subsequently found to have been at that time *thirty miles* distant, and *seventeen miles beyond* the horizon.

MIRAGE.

442. During the late Exploring Expedition, a singular instance of mirage was seen off Terra del Fuego, from the decks of the Vincennes and Peacock, and which is thus related. "On the 17th of February, 1839, we had an extraordinary degree of mirage or refraction of the Peacock, exhibiting *three* images, two of which were upright and *one inverted*. They were all extremely well defined. The temperature on deck was 54° Fah., that at the mast-head 62° Fah. A vessel, that was not in sight from the Vincennes' deck, became visible, and the land was much distorted, both *vertically* and *horizontally*.

Relate the several cases of mirage, ¶ 440—445.

On board the Peacock, similar appearances were observed of the Vincennes and Porpoise. There was, however, a greater difference between the mast-head temperature and that on deck, the thermometer standing at 62° Fah. at the mast-head, while on the deck it was but 50° Fah., being a difference of 12°; that on board the Vincennes differed only 8°."

443. Simpson, while exploring the coasts of the north polar seas, in the summer of 1837, beheld a remarkable display of the mirage. As he rowed over the tranquil ocean, he seemed to be traversing a *valley;* the waters apparently *rising* on either hand, like the *sides of a mountain*, and the huge icebergs upon their surface appearing ready to topple down upon him.

444. During the march of the French army over the sandy plains of Egypt, many instances of the *mirage* occurred. The villages, situated upon small eminences, were successively seen like so many islands in the midst of an *extensive lake*, and beneath each village appeared its *inverted image*. In the same direction, an image of the blue sky was seen, clothing the sand with its own bright hues, and causing the wilderness to appear like a rich and luxuriant country. So complete was the deception, that the troops hastened forward to refresh themselves amid these cool retreats; but, as they advanced, the illusion vanished, only to re-appear at the villages beyond.

445. This phenomenon is so common on the deserts of Asia and Africa, that the Koran calls every thing deceitful by the word *serab*, which signifies mirage. It remarks, for example, that "the actions of the incredulous are like the *serab* of the plain; he who is thirsty takes it for water, and finds it to be nothing."

446. While Baron Humboldt was at Cumana, he frequently saw the islands of Picuita and Boracha, apparently *hanging in the air*, and sometimes with inverted images; and at Mesa de Pavona, cows were beheld

---

Where does this phenomenon frequently occur?
What instances are given by Humboldt and Tschudi?

seemingly suspended in the air, at the distance of 2,132 yards.

When Dr. Tschudi and his party were traversing a deep sandy plain, near the river Pasamayo in Peru, they beheld the figures of themselves, *riding over their own heads, magnified to gigantic proportions.*

447. FATA MORGANA. This name is given to an extraordinary optical phenomenon, which has been often seen in the straits of Messina, between the island of Sicily and the Italian coast. It has been described by many writers, and, though known for centuries, has but lately been considered as the effect of mirage. The following is the description by Antonio Minasi, which is regarded as the most correct.

" When the rising sun shines from a point, whence its incident ray forms an angle of about 45° on the sea of Reggio, and the bright surface of the water in the bay is not disturbed either by the wind or the current, a spectator placed on an eminence in the city of Reggio, with his back to the sun, and his face to the sea, suddenly beholds in the water numberless series of pilasters, arches, castles well delineated, regular columns, lofty towers, superb palaces, with balconies and windows, extended valleys of trees, delightful plains with herds and flocks, armies of men on foot and horseback, all passing rapidly in succession along the surface of the sea."

In a peculiar state of the atmosphere, when its dense vapors extend like a curtain over the waters, the same objects are not only reflected from the surface of the sea, but are likewise seen in the air, though not so distinct or well defined, and if the atmosphere is slightly hazy, the images seen upon the surface of the water are *vividly colored* or fringed with all the prismatic hues.

448. But a most extraordinary instance of the mirage occurred at Hastings, on the coast of Sussex, on the 26th of July, 1798. The cliffs of the French coast are *fifty miles distant* from this town, and in the usual state of the atmosphere, are *below the horizon* and completely

---

Describe the Fata Morgana.

hid from view; but on the day mentioned, at five o'clock P. M., they were seen extending to the right and left for *several leagues,* and apparently only *a few miles off.* As the narrator, Mr. Latham, walked along the shore, the sailors, who accompanied him, pointed out and named the different places on the opposite coast, which they were accustomed to visit. By the aid of a telescope, small vessels were plainly seen at anchor in the French harbors, and the buildings on the heights beyond were distinctly visible.

The Cape of Dungeness, which at the distance of 16 miles from Hastings, extends nearly two miles into the sea, appeared quite close to the town, and the fishing boats, that were sailing at the time between the two places, were magnified to a high degree. This curious phenomenon continued in its greatest beauty for more than *three hours.* The day was extremely hot, without a breath of wind.

449. A remarkable mirage of Dover Castle, was seen by Dr. Vince and another gentleman, on the 6th day of August, 1806, at Ramsgate.

Fig. 27.

MIRAGE—DOVER CASTLE.

The summits, $v\ x\ w\ y$, of the four turrets of the castle, (fig. 27.,) are usually seen *beyond the hill* A B, which lies between the castle and Ramsgate; but, on this day not only the turrets were visible, but the whole castle, $m\ n\ r\ s$, appeared as if it were on *the side of the hill next to Ramsgate.*

Relate the account of the mirage at Hastings, and of that at Ramsgate.

Between the observers and the shore, from which the hill rises, there was about six miles of sea, and from thence to the top of the hill the distance was about the same. Their own height above the water was nearly seventy feet.

450. ORIGIN. The cause of mirage has been partially stated; but the subject demands a more complete explanation. The phenomena may be divided into three classes, viz.: those produced by *refraction*, those produced by *refraction* and *reflection conjointly*, and those produced by *reflection only*.

451. The image of Dover Castle was probably produced by refraction, simply; for the atmosphere gradually increasing in density from the lofty heights of the castle to the level of the sea, the rays of light proceeding from the edifice, reached the eyes of the spectators in a curved line, like those which emanate from a star, (Art. 438;) and the whole structure therefore appeared to the observers above its true position.

452. Phenomena, like those observed by Scoresby, are attributed to the combined influence of refraction and reflection. At such times, the stratum of air in contact with the sea is *colder* than that immediately above (Art. 442), and this likewise colder than the next superior stratum, and so on. Consequently, to a certain extent, the *density* of the atmosphere *decreases with the distance from the ocean*, and, under these circumstances, the rays of light from a ship may be so changed in direction, as they proceed through the air, that the observer will behold both *erect* and *inverted images above* the real object.

453. ERECT AND INVERTED IMAGES ABOVE THE OBJECT. The annexed figure will aid us in perceiving how *erect* images are caused.

What is said respecting the cause of mirage?
Into what classes may the phenomena be divided?
Explain the mirage of Dover Castle.
To what is attributed the phenomena of erect and inverted images above the object?
What is the state of the atmosphere as regards temperature at such times?

OPTICAL PHENOMENA.

Fig. 28.

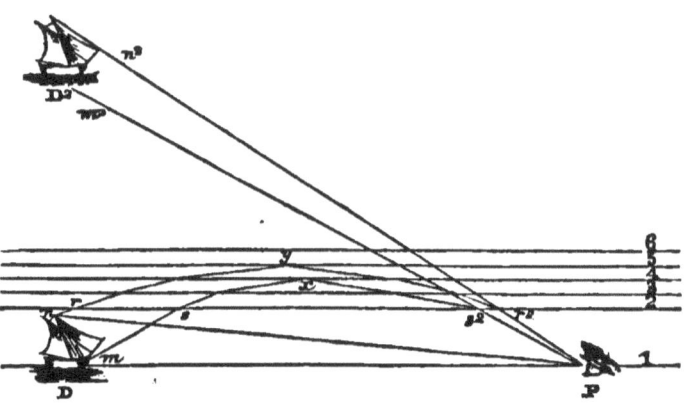

ERECT IMAGE ABOVE THE OBJECT.

Let D be a ship, seen in the horizon in its true position, by the direct rays n P, m P, coming to the eye at P, through the stratum of air of uniform density, lying between the eye and the ship. Let 1–2, 2–3, 3–4, &c., be parallel strata of the atmosphere, decreasing in density from 2 to 6; and $n\,r$ and $m\,s$, rays of light, proceeding upwards from the top and bottom of the ship. As these rays at r and s pass from the first stratum into the second, which is rarer, they are *bent downwards*, or from the perpendicular, according to a well-known law of optics, (C. 706,) and this change in direction continually occurs as they pass successively into strata still more and more rare; until at last, as at x and y, they meet the next superior stratum so obliquely, that they are unable to enter it, and are then totally reflected from the lower surfaces of strata 4 and 5, at the points x and y.

The rays, on their return, are now *refracted downwards*, or towards the perpendicular, (C. 706,) in passing from the *rarer* into the *denser media*, and converge to the eye at P, which sees the vessel in the direction of the *last refracted rays*. The ship D is therefore beheld at D^2 by the rays $P r^2\,n^2$, and $P s^2\,m^2$.

Explain from figure 28. the phenomenon of an *erect* image above the object.

454. In figure 28., the upper ray *before* reflection is the upper ray *after* reflection, and the image consequently appears *erect;* but if, as in figure 29., the rays *cross each*

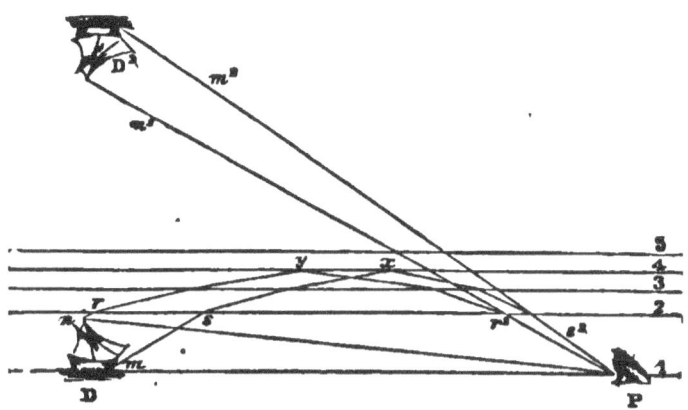

Fig. 29.

INVERTED IMAGE ABOVE THE OBJECT.

other before they reach the eye at P, then the image will appear *inverted,* as is evident from the inspection of the figure.

Under peculiar circumstances it may happen, that of *two sets of rays,* one from the top and the other from the bottom of an object, some may cross each other before they meet the eye and some may not; and then both *erect and inverted images* will be seen at the *same time.*

455. MAGNIFIED IMAGES. The real object in figures 28. and 29., is seen through the horizontal strata, under the visual angle n P m. If n^2 P m^2, the angle under which the image is seen, is *greater* than n P m, the image will be *magnified* (C. 746) in the direction of its *length;* and if an *increase* of the *lateral* visual angle occurs at the *same time,* then the image will be likewise *magni-*

Explain from figure 29. the phenomenon of an inverted image above the object.
When may *both images* be seen at the same time?
When may the image be magnified both *vertically* and *horizontally?*

186 OPTICAL PHENOMENA.

fied in breadth, and will appear as if seen through a telescope.

The mirage at Hastings was probably due to this cause.

That such a lateral displacement is possible, is evident from the remarkable mirage beheld by Messrs. Soret and Jurine, on the Lake of Geneva, in Sept. 1818, and which is shown in figure 30.

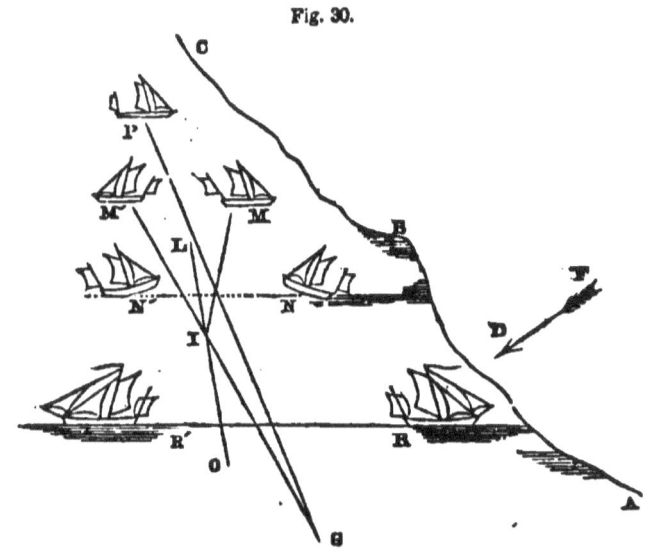

Fig. 30.

LATERAL MIRAGE.

456. The curve A B C represents the east bank of the lake. A boat, with all her sails set, was at P, advancing towards Geneva, and was seen, by the aid of a telescope, in the direction of G P, from Jurine's house, at the distance of six miles. As the boat successively occupied the positions M N R, a *lateral image* was clearly seen at the corresponding points M' N' R', approaching with the boat, but appearing to recede to the *left* of G P, while the boat receded to the *right*. When the sun shone full upon the sails, the image was visible to the naked eye.

Describe the lateral mirage seen at Geneva (figure 30.), and explain its cause.

The direction of the sun's rays at the time of observation, is shown by the arrow F D.

457. This curious phenomenon is supposed to have been caused in the following manner. The air at M N R, had been in the *shade* all the morning, while that at M' N' R' had been *warmed by the sun;* the two portions therefore were of *different densities*, and the surface which separated the warm air from the cold was probably *vertical.* The rays of light proceeding from the boat might, in this case, fall upon the *vertical surface* of the warm stratum as upon a *mirror,* (Art. 461,) and being thence reflected to the eye of the observer, an image of the boat would appear *behind this surface.* Thus, if L I O was a part of this surface, the ray M I from the boat at M might be reflected from it in the direction I G, and an image would then be seen by an observer at G, in the direction G I M'.

458. IMAGES BELOW THE OBJECT. Illusions, like those which were seen by the French army, arise from a condition of the atmosphere exactly the reverse of that which occasions the images to appear above the object.

Upon the arid plains of Asia and Africa, when the sand is intensely heated, the temperature of the air decreases and its *density increases from the surface upwards* to a certain height, where it is nearly uniform. A ray of light, therefore, proceeding from an elevated object towards the ground, must necessarily pass through strata of decreasing density, and will consequently, upon principles already explained, after a number of refractions be *reflected upwards*, causing *images* to be seen *below their objects.*

459. For the sake of illustration, suppose (fig. 31.) that 1-2, 2-3, 3-4, 4-5, &c., are atmospheric strata *decreasing in density* from the height, 6, to the surface of the ground at 1. Let D represent a palm tree, which is seen in its true position by the eye at P, through an atmosphere of uniform density. The rays of light *e a*,

---

Under what circumstances are images seen *below* the object?
What is then the state of the atmosphere, in respect to temperature?

188 OPTICAL PHENOMENA.

Fig. 31.

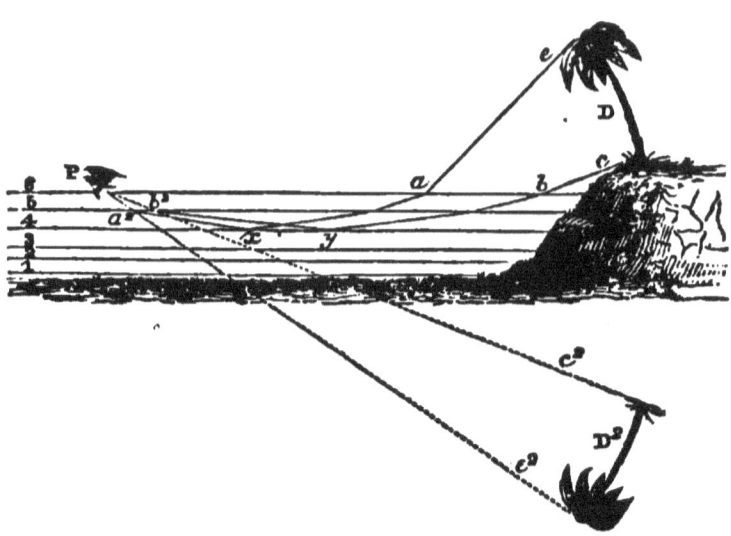

INVERTED IMAGE BELOW THE OBJECT.

$c\,b$, which proceed from the top and bottom of the tree, in passing successively from *denser* into *rarer media*, will be constantly *bent upwards*, until at last they suffer *total reflection* at y and x; and then, *crossing each other*, are again *refracted* through the upper strata converging to the eye at P. An inverted image D^2 will therefore be seen *below* the *real object* in the direction of the *last refracted rays* P $b^2\,c^2$ and P $a^2\,e^2$.

460. The observer is led to imagine these images in the midst of a *lake* from the circumstance, that the ascending currents of warm air, mixing with the colder strata, impart a *tremulous motion* to the images seen through them; and thus they appear to be agitated, as if floating upon a slightly ruffled surface. A difference of *three or four degrees* in temperature is sufficient to occasion appearances of this kind.

461. IMAGES PRODUCED BY REFLECTION. It is probable that the mirage is sometimes produced by *reflection only*, as from a *plane mirror*, and the instance witnessed

---

Illustrate from figure 31.

by Capt. Mundy when travelling in India, and thus related in his Journal, may have proceeded from this cause. "A deep, precipitous valley below us, at the bottom of which I had seen one or two miserable villages in the morning, bore in the evening a complete resemblance to a beautiful *lake*; the vapor which played the part of water ascending nearly half way up the sides of the vale, and on its bright surface trees and rocks being distinctly *reflected*. I had not been long contemplating this phenomenon, before a sudden storm came on, and dropped a curtain of clouds over the scene."

The *fata morgana* is attributed to the *reflection* of the rays of light from the surface of the sea and vapor.

462. The reflecting surface of a stratum of air may possibly at times possess a *concave form*, so as to present a *magnified image* of the object. (C. 732.) In this way, the gigantic images seen by Dr. Tschudi were probably produced; the reflecting surface of the stratum of air being nearly *vertical*.

463. Another instance of this kind occurred, during the last war with England, when Commodore Hardy was lying off Boston. A figure of a sailor of a *colossal* size, was seen by his whole ship's crew *reflected* in the heavens, during a peculiar state of the atmosphere.

464. To the same cause must be attributed the extraordinary phenomenon, which occurred in the parish of Migne in France, on the 17th of December, 1826. It was Sunday, and 3000 persons were engaged in the exercises of the Jubilee. As a part of the ceremony, a *large red cross*, twenty-five feet high, was planted beside the church, in the open air. Towards the close of the day, while one of the clergy was addressing the multitude, and reminding them of the miraculous cross beheld in the sky by Constantine and his army, a *cross* was seen at that moment *in the heavens*, directly before the porch of the church, and at the height of two hun-

---

Explain in what manner images are produced by reflection.
Are they ever magnified?
How is this accounted for?
Give instances.

dred feet above the ground. Its length was nearly one hundred and forty feet, its breadth from three to four, and it shone with a bright silvery hue, *tinged with red.*

The assembled multitude were struck with awe, many regarding it as a miracle, and such was the extraordinary sensation produced throughout the country, that a committee was appointed to investigate this phenomenon.

From the circumstances detailed in their report it is evident, that the cross in the sky was the *magnified image* of the cross before the church, and *reflected* from the *concave surface* of some *atmospheric mirror.* The image possessed exactly the shape and proportions of the wooden cross, it was *tinged* with the *same color*, and the state of the air at the time was favorable to the formation of such images.

465. SPECTRE OF THE BROCKEN. The gigantic spectre which is supposed to haunt the Hartz mountains in Hanover, and is seen at sunrise from the Brocken, the loftiest peak of the range, is produced in a different manner. It is in fact nothing more than the *shadow of the observer*, cast upon the thin vapors then floating in the sky. The cut below represents the spectre, as seen by Mr. Haue and another person, on the 23d of May, 1797.

Fig. 32.

SPECTRE OF THE BROCKEN.

Standing on the summit of the Brocken at sun-

How is the spectre of the Brocken explained?

they at first beheld upon the transparent vapors opposite to the sun, *two human figures of immense size* which *imitated* all their gestures. In a short time they vanished, but soon re-appeared, and were joined by a *third*, which likewise *mimicked every motion and attitude* of the observers. Similar phantoms are beheld over the lake of Killarney, in Ireland.

A spectacle of this kind was observed by Baron Gros, Secretary of the French Legation in Mexico, during his ascent of Popocatepetl, in April, 1834. When he had attained a very great height, he distinctly beheld, on the morning of the 29th at sunrise, the *shadow of the entire volcano cast upon the atmosphere*. It appeared as an *immense circle of shade*, through which the whole country below could be plainly seen, and was bounded by a vapor moving from north to south. As the sun rose the shadow descended, becoming more and more transparent, and in the space of two or three minutes was entirely dispersed.

466. ARTIFICIAL MIRAGE. The phenomena of the mirage have been artificially produced by Dr. Brewster, in the following manner.

Fig. 33.

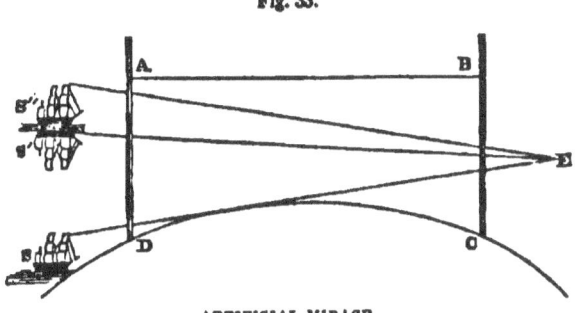

ARTIFICIAL MIRAGE.

A B C D, figure 33., represents a trough with glass sides, A D, C B, and filled with water up to the level A B. If a hot iron is held near the surface of the water the heat descends, and a change takes place in the den-

---

Describe this phenomenon.
What phenomenon was beheld by Baron Gros?
In what manner may an artificial mirage be produced?

sity of the fluid, the *density increasing from the surface to the bottom.* When the heat has almost reached the bottom, if a small object, as a toy-ship, is then placed at S, the eye at E will see an *inverted image* of the ship at S' and an *erect* one at S": an appearance similar to the mirage observed by Dr. Vince.

CHAPTER IV.

OF CORONAS AND HALOES.

467. CORONAS. When light, gauze-like clouds float before the sun and moon, their disks are sometimes seen encircled by one or more colored rings, which are termed *coronas* or *crowns.* This appearance is more frequently beheld about the moon; for the eye is usually too much dazzled by the brilliancy of the sun to discern the hues that surround its orb. To observe them with distinctness, they must be viewed by reflection from a blackened mirror, which tempers the vividness of the solar rays.

468. If the coronas are complete, the rings are each composed of several concentric circles: the first, counting from the disk, is of a *deep blue,* the second *white,* and the third *red.*

These three circles constitute the *first* ring. In the *second* the order of colored circles, reckoning the same way, is *purple, blue, green, pale yellow* and *red.*

469. Rarely, however, are coronas thus perfect; for more often blue mingled with white is observed near the disk; this is followed by a red ring, its inner margin clearly defined, but its outer limit blended with the

---

Of what does chapter fourth treat?
What are coronas?
Around what orb are they most frequently seen?
What is the appearance of coronas when complete?
Are they usually perfect?

succeeding circles. If beyond this, a second red ring is seen, a green circle occupies the interval between the two. Kaemtz discovered, that the *distance* of this latter circle from the centre of the sun varied, according to the state of the clouds and atmosphere, from *one* to *four degrees.*

470. ORIGIN. According to Fraunhofer, coronas are caused by the *diffraction of light;* by which is understood *the change that a ray of light undergoes in passing across the edge of an interposed body,* in consequence of which it is decomposed into the seven primary colors, as if transmitted through a prism.

471. This effect may be seen if a ray of light is admitted into a darkened room, through a very small opening, as a pin-hole, and a knife-blade placed *across* the ray. If the shadow of the blade is now received upon a white screen, *fringes of colored light* will be observed on each side of the shadow, arranged in *the order of the prismatic hues,* commencing with the *blue* and terminating with the *red.* The manner in which coronas are produced by diffraction, requires a somewhat extended explanation. If we cut a fine slit with a knife in a card, and view through the opening any luminous object, as a candle, we shall perceive on each side of the aperture a *row of colored images,* which are those of the candle. Each of these images possesses all the colors of the spectrum, and the order of their tints is the same, the *violet* being *nearest* to the *aperture,* and the *red* the *most distant* from it.

472. The images will be more *numerous* and *brilliant,* if instead of a *single* opening, a system of *many* apertures is arranged, equal in size, and equally distant from each other. This may be effected by ruling upon a piece of glass with the point of a diamond, *fine, parallel,* and *equi-distant lines, several hundreds in an inch.* If in a darkened room we look through a plate of glass thus

---

What is the opinion of Fraunhofer respecting the cause of coronas?
Explain *diffraction.*
How may this phenomenon be exhibited?

prepared, at a small opening in the window-shutter through which the sun-light comes, a *row of prismatic images* of this opening will be seen on each side of it; their direction being *perpendicular* to the *lines*.

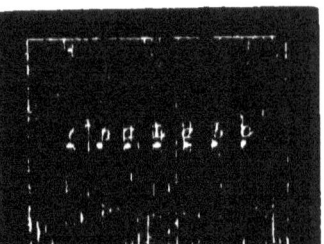
Fig. 34.
PRISMATIC IMAGES.

Thus, if figure 34. represents such a plate of glass, the *white lines* being the *ruled lines*, and the aperture is viewed through the point, L, a row of images will be seen on each side of L at the corresponding points, *a a*, *b b*, *c c*, and in a line *perpendicular* to the ruled lines. The tints are in the same order as in the first experiment, and are caused by the decomposition which the rays of light undergo in passing by the edges of the unruled intervals of glass.

If another series of parallel lines is drawn at *right angles* to the first; a *second row* of colored images will start up in a direction *perpendicular* to that of the *first images;* and, if the ruled lines in both series are equally distant from each other, the first set of images *a a a a*, will all be situated at the same distance from L, *each one* on the *middle* of the *side of a square*, whose centre is L. The same will be true of the *second set b b b b*, and so of the rest.

If there were *three series* of ruled lines, equally inclined to each other, they would form regular *six-sided figures* or *hexagons,* and the images would be found on the sides of *hexagons* as in figure 35., where the white dots represent the places of the images.

473. Now it is evident, the number of series may be so increased, that the figures formed by the lines shall have so many sides as not to differ essentially from *circles ;* and then the colored images would *touch each*

Show, by the aid of figures 34., 35. and 36., the peculiar manner in which diffraction operates so as to produce coronas.

OF CORONAS AND HALOES. 195

other, forming *rings of prismatic images* around the luminous point, the *blue* being the *innermost* color, and the *red* the *outermost*. Thus, for instance, if in figure 35., the number of series was so multiplied, the *largest hexagon* would be changed into the *limiting circle*, and the images, *a a*, *b b*, *c c*, &c., would form prismatic circles around L as a *common centre*.

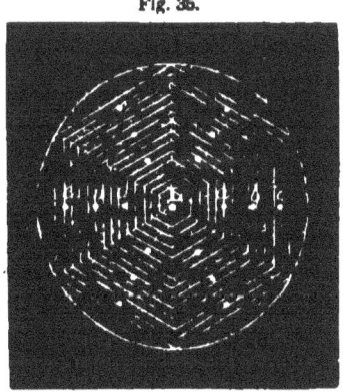

Fig. 35.

PRISMATIC IMAGES.

474. These rings appear in great beauty when a luminous object is viewed through a plate of ruled glass, the lines forming concentric circles, many hundreds, and even thousands being contained in an inch; the *distance*, *number*, and *brilliancy* of the rings increasing with the *fineness* of the *lines*, and the *narrowness* of the *transparent intervals*.

475. Similar prismatic rings are beheld, whenever the transparent intervals through which the object is seen are grouped symmetrically around a point. Thus Fraunhofer, in looking at a luminous object through a number of small glass balls of equal size, placed between two parallel plates of glass, saw it surrounded by several colored circles or coronas. Nor is this surprising, for the apertures between the balls through which the light came, are arranged concentrically around a point, like the transparent intervals in the case of the circularly ruled lines. This will be seen by a reference to figure 36., where the *dark circles* represent the *balls*, and the *light parts* between them the *intervals* through which the rays come from the luminous object, which is supposed to be situated behind the figure. Now it is obvious at a glance, if the eye is fixed upon B, that the intervals lie in the *circumferences* of circles, whose com-

mon centre is the *middle point* of B. If we look at A, the intervals are arranged around this ball in a similar way, and so of any other ball. In viewing, therefore, a bright object through the balls, it should exhibit the same appearances as if seen through the circularly ruled glass, and this is found to be the case.

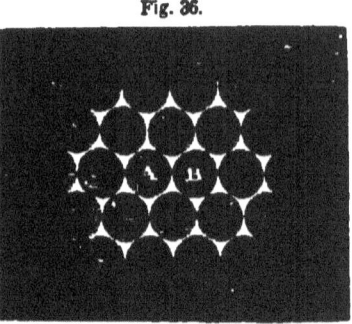

Fig. 36.

GLASS BALLS.

476. Now the *globules of vapor*, of which fogs and clouds are composed, are arranged in a similar manner throughout the atmosphere, and act upon the light of the sun and moon as if they were so many small glass balls. When, therefore, the rays of the moon, for instance, reach the eye of the observer, after passing between the particles of light, interposing vapors, he will often see her orb surrounded by beautiful *coronas*, glowing with the rich colors of the spectrum.

477. Coronas are only seen when the globules of vapor are comparatively *few*, and are of *equal size*. If they are *too numerous* a dense cloud is formed, and the intervals being closed by the globules, no rays can pass through them. If they are *few* in *number* but *differ in size*, then the *intervals* are not *symmetrically arranged*, and the sun or moon will appear surrounded by *a glory*, or bright circle of white light.

The distance of coronas from the luminous body is not always the same. The *smaller* the *particles*, the *greater* is the *diameter* of the rings.

478. When white clouds, having the form of the cirro-cumulus, float near the sun, bright, prismatic colors are often seen, by the aid of a blackened mirror,

---

When are coronas only seen?
What is the result when the particles of vapor differ in size?
Why will the rings vary in magnitude?
What is said respecting the edges of cirro-cumulus and cumulus clouds, when passing near the sun and moon?

fringing the edges that are parallel to the horizon. The fringes are generally *green* within, bordered by *two red lines*.

If the air is pure, and the moon shines brightly, the light and broken edges of cumulus clouds, as they pass near her disk, are sometimes seen in like manner fringed with prismatic hues, the *purple* tint being the *nearest* color, and *the red* the *most distant*.

479. ANTHELIA. Anthelia are coronas seen by *reflection*, when the back of the observer is towards the sun ; and are so called from the Greek words *anti*, opposite, and *helios*, the sun.

If the plain surface of the circularly ruled glass (Art. 474) is blackened, and the luminous object seen by *reflection* upon the ruled side, its image will appear surrounded by colored rings precisely like those that encircle the object itself, when viewed, as in the first case, by *transmitted* light.

In analogy to this, when the shadow of a person is cast upon a stratum of vapor, the head of the observer, under favorable circumstances, is seen *surrounded* with *prismatic circles*.

480. A beautiful display of this kind was witnessed from the summit of Mount Lafayette, fifteen miles from Mount Washington, on the 7th of August, 1826. In the afternoon of the day in question, two gentlemen were standing upon this lofty eminence, a thunderstorm was raging beneath them, and a sea of vapor shut out the vales from view. A light mist was at this time falling, when suddenly the sun burst through the clouds above, and the observers saw their *shadows resting upon the vapor before them*, their *heads surrounded* with brilliant, *prismatic rings*. The circles were apparently *ten* or *twelve* feet in *diameter*, perfectly defined, and their tints were exceedingly rich and vivid. This phenomenon lasted for the space of twelve or fifteen minutes, when it gradually vanished.

481. In the polar seas, when the stratum of fog that

Define anthelia. Relate the instances given.

rests upon the ocean rises to the height of about three hundred feet, a person, stationed upon the mast of a ship, eighty or a hundred feet above the water, perceives in the fog opposite the sun, one or more circles around the shadow of his head. They are all concentric; their *common* centre being in the *imaginary line drawn from the sun through the eye of the spectator to the fog beyond him.* The number of circles varies from *one* to *five*, and when the sun is bright, or the fog thick and low, they are usually numerous and highly colored.

482. On the 23d of July, 1821, Scoresby saw four concentric circles around his head, with the series of colors arranged in the following order:

1st circle, *white, yellow, red, purple.*
2d circle, *blue, green, yellow, red, purple.*
3d circle, *green, whitish, yellowish, red, purple.*
4th circle, *greenish white.*

The colors of the *first* and *second* rings were very brilliant, those of the *third* faint, and only seen at intervals, while the *fourth* exhibited only a slight tinge of green. According to Scoresby, anthelia are always seen in the polar regions whenever *fog* and *sunshine* occur at the *same time.*

483. Several philosophers have supposed that anthelia, or coronas opposite to the sun, are caused by the passage of light through frozen particles of vapor, but this phenomenon has frequently occurred, when the temperature of the air was so high as to preclude this idea.

Thus Kaemtz often beheld anthelia upon the Alps, when the temperature of the air was 50° Fah., at a short distance from the fog. Their explanation upon the principle of diffraction is the most satisfactory, and the truth of this theory is strongly confirmed by an observation of Kaemtz, who, on one occasion, first saw a *corona* when the cloud was *between* himself and the sun,

---

What is the opinion of some philosophers in regard to anthelia?
What objection can be urged against this view?
| What fact is stated by Kaemtz?

and then an *anthelion* from the *same* cloud when it was *opposite* to the *sun*.

HALOES.

484. *Haloes are circles of prismatic colors* about the *sun* and *moon;* they differ from coronas in three particulars; first, their structure is often more complicated; secondly, their diameter is greater; and thirdly, the order of colors is reversed, the *red* being *nearest* the luminary.

485. The several parts of this phenomenon may be thus classified, 1st, *Circles surrounding the orb which occupies their centre.* 2d, *Circles passing through the orb.* 3d, *Arcs of circles touching those of the first class.* 4th, *Parhelia and paraselenæ* or *mock-suns* and *mock moons,* found at the points where the circles cross each other.

486. FACTS. The annexed figure represents a halo around the sun, observed by Scheiner, in 1630. In the cut, S is the sun, A B C a circle about 45° in diameter, and D E F another circle, its diameter being nearly 95° 20', the *sun being in their common centre.* Both the circles were colored like the primary rainbow, but the red was next to the sun, the other colors succeeding in the natural order. D S F is a third whitish circle *passing through the centre of the sun*, and H E G a portion of a fourth *touching* D E F at E. At A, C, D, and F, were mock-suns; the same phenomena were seen at B

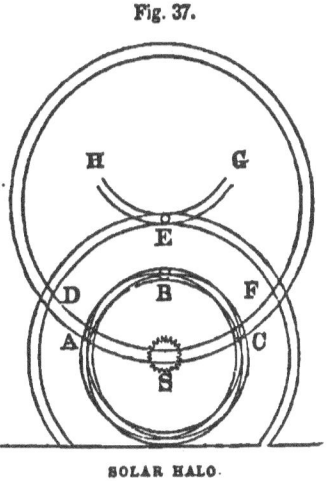

Fig. 37.

SOLAR HALO.

What are haloes?
How do they differ from coronas or crowns?
How are the several parts of the halo classified?
Describe the three haloes recorded.

and E. The mock-suns, A and C, were of a *purplish red* next to the sun, while D and F were *entirely white*, the former were also more brilliant, continuing visible for three hours together, while the light of the latter was faint and fluctuating.

The mock-suns B and E were almost the first to appear and the last to fade, excepting A, and throughout the whole phenomenon, which lasted five hours, they were perpetually changing in *magnitude* and *color*. B was formed in a peculiar manner, the halo A B C was composed of several *intersecting* circles, and at one of these intersections the mock-sun B appeared.

487. On the 9th of Sept. 1844, a halo of a somewhat complicated structure was seen by many observers, both at New Haven and at Hartford, Ct. It continued for the space of *four hours*, commencing about 10 A. M. and ending at 2 P. M.

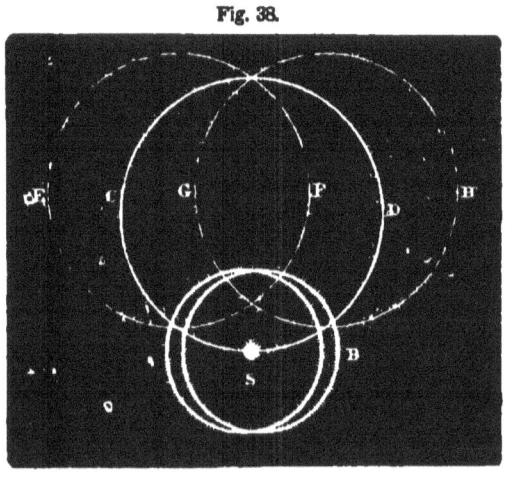

Fig. 38.

SOLAR HALO.

Its appearance is shown in figure 38., where S represents the sun, A B the ordinary halo of about 45° in diameter, and C D a circle whose *centre* was in the *zenith* while its circumference passed through the sun. Directly north of the zenith, upon the circumference of C D, a *parhelion* appeared at the intersection of C D with the circles E F and G H. which were both equal in size to itself.

The halo A B exhibited at times *bright prismatic tints*, and was attended by an ellipse or oval, as seen in the figure. The other circles were *white*, and fainter according as they were situated farther from the sun.

488. On the 30th of March 1660, at Dantzic, Hevelius beheld, about one o'clock in the morning, the halo shown in figure 39. When first perceived the moon at M was surrounded by a complete whitish circle, A B C, 45° in diameter, while at A and C were two *mock-moons* displaying *various colors*, and shooting out at intervals very long and whitish streams of light. At two o'clock the larger circle, D E F, was seen, reaching down to the horizon, having a diameter of 90°.

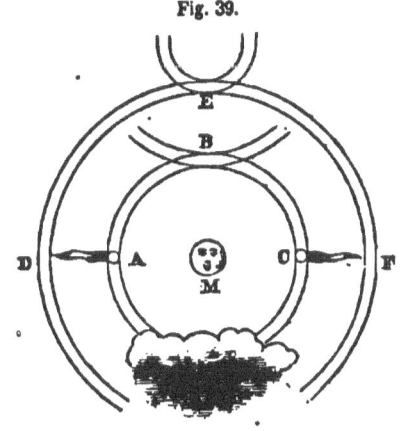

Fig. 39.

LUNAR HALO.

The tops of both circles were *touched* by *colored arches*, like inverted rainbows, the *red tint* being *next to the moon*. The arch at B was a part of a circle equal in size to D E F, while that at E was a portion of a circle of the same magnitude as A B C.

489. Such is the general structure of haloes, and the identity that exists in the magnitude and arrangement of the several parts clearly shows, that they must originate in certain fixed laws; but what those laws are has not yet been fully determined.

490. ORDINARY HALO OF 45°. The most satisfactory explanation of this halo is that given by Mariotte and Dr. Young, who suppose it to be caused by *the refraction of the sun's rays, as they pass through crystals of frozen vapor, floating in the upper regions of the atmosphere.*

What does the general structure of haloes indicate?
Explain the origin of the halo of forty-five degrees.

491. For the sake of illustration we will suppose that A, figure 40., is a crystal of ice or snow, having its refracting angle equal to 60°, which is the usual angle of such crystals, and that S E and S P are parallel sunbeams, and E the eye of the observer. The ray, S P, passes through the crystal as through a *prism,* and is decomposed into its original colors, the *greatest amount* of *prismatic light* reaching the eye when the *angle of deviation,* S E R, is *about twenty-two degrees and a half.*

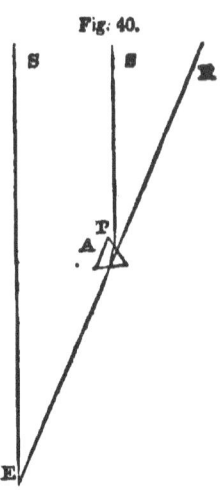

REFRACTION THROUGH ICE-CRYSTALS.

492. Now, it is well known, that in cold weather the air near the earth is often filled with fine needle-shaped crystals of ice, and that in the higher regions of the atmosphere, above the limit of perpetual congelation, crystalized vapor exists in summer as well as in winter. (Art. 237.)

493. If we then suppose a stratum of ice-crystals floating in the air so thin that the sun is distinctly seen through it, though veiled as by a slight mist; an observer will behold this luminary encircled by *rings of colored light,* proceeding from those crystals whose angular distance from the sun is about *twenty-two degrees and a half.*

The diameter of this circle or halo, will of course be nearly 45°, and the *red tint* will be *next* to the sun since it suffers *less refraction* than the *blue.*

494. It might be objected, that the crystals of snow, when floating in the air, would not *naturally* assume such positions as to refract the light properly to the eye; but it can be proved by rigorous calculations, that if the vast number of crystals which compose the stratum, take *every possible position, one-half* of the sun-

---

How is the objection answered, that the crystals of ice would not naturally assume such a position as to refract the light to the eye?

light will pass through them; and that *one-third* of the transmitted rays will reach the eye within a *range of one and a half degrees*, viz., when the *angle of deviation* S E R varies from 21° 50' to 23° 22'.

Such is the theory in regard to the origin of the *ordinary halo*, and the probability of its truth is strengthened by the fact, that fine crystals of ice are known to produce curves and circles of prismatic light.

495. On the 23d of March, 1845, Prof. Snell, of Amherst College, beheld a most beautiful phenomenon. As he stood facing the sun, which had just arisen, he observed upon the dead grass before him a *curved, horizontal band* of light, *three* or *four feet broad*, glowing with all the colors of the rainbow. The top of the curve was twelve or fifteen feet distant, while the two branches extended several rods to the right and left. The long spires of dead grass were fringed with *frost-crystals*, and the cause of this brilliant arch was justly attributed to the refraction of the sun's rays as they traversed these minute prisms.

496. If a distant light, as a street-lamp, is viewed through a pane of glass upon which the vapor of a room has crystalized, two or more fine *haloes* will be distinctly seen surrounding the luminous object. The same appearances are presented to the eye if we substitute a plate of glass upon which a few drops of a saturated solution of alum have rapidly crystalized.

497. Extraordinary Halo of 90°. The halo of *ninety degrees* is also supposed to be owing to the refraction of light through crystals of ice or snow; the crystals being six-sided prisms. (Art. 283.)

498. Circles passing through the Sun. These are often *highly colored*, and when the sun is near the horizon, a portion of a vertical circle sometimes presents the appearance of an *upright, luminous column.*

What facts are stated to show that minute ice-crystals can produce haloes?
How is the halo of ninety degrees caused?
What is said respecting the circles passing through the sun?

Many years ago, on a very cold morning, there were seen at West Point, above the sun, vertical columns of light of exceeding splendor, tinted with all the *prismatic colors*, and surpassing in brilliancy the hues of the rainbow. A similar phenomenon was observed at the same place, by Prof. Twining, on the 5th of Jan. 1835; but the prismatic tints were wanting.

On the 2d of January, 1586, an extraordinary display of this kind was seen by Roth, at Cassel. Before the sun rose, an *upright column of light* illumined the sky, at the point where the sun was about to appear. Its breadth was equal to that of the sun, and it glowed like a vivid flame. An *image* of the sun next appeared so brilliant as to be taken for the orb itself. This was immediately followed by the *true sun*, which was directly succeeded by a *second image*. The luminous column with its *three suns* was visible for the space of an *hour;* the three suns were exactly similar, only the *true one* was the *brightest*.

499. A similar phenomenon was observed on the 21st of February, 1847, by Lieut. Abert and his party, during their exploration of New Mexico, and is thus described by Abert in the official report of the expedition:

"The snow had heaped up around the rest of the tents so that the inmates were obliged to desert them, and take refuge in the wagons. About mine, the wind had swept in such a way as to keep open a path around it, although the snow was on a level with the ridge-pole of the tent. We now broke up some boards that were in the wagons, and kindled a little fire. Soon the sun rose; but, instead of *one* sun, we had *three;* all seemed of equal brilliancy; but, as they continued to rise, the middle one only retained its circular form, while the others shot into *huge columns of fire*, which blended with the air near their summits. The breadth of the columns was that of the sun's apparent diameter, and their height about twelve times the same diameter; they were between twenty and thirty degrees distant from the sun. Before the sun had risen more than ten degrees the phenomenon entirely disappeared."

500. The origin of these circles, as well as of those which belong to the *third class*, is attributed to the action of snow-crystals upon the rays of light, and philosophers have discovered much ingenuity in framing hypotheses to account for these phenomena.

501. PARHELIA AND PARASELENÆ. The images of the *sun* observed in haloes, are called *parhelia*, from the Greek words *para*, near, and *helios*, the sun ; while those of the *moon* are termed *paraselenæ*, from *para*, near, and *seléna*, the moon : they have also received the names of *mock-suns*, and *mock-moons*. These images are found at the intersection of the different circles, and are formed by the accumulation of light at these points. That such an increase of light occurs is obvious: for if two equally bright circles cut each other, the place where they cross will be *twice* as bright as the circles themselves. The parhelia and paraselenæ are tinged with the colors of the ordinary halo, and have frequently appended to them a waving stream of light.

What is said as to their origin ?
What are parhelia and paraselenæ ?
In what manner are they produced ?

PART VI.

LUMINOUS PHENOMENA.

CHAPTER I.

OF METEORITES.

502. METEORITES are those *solid fiery bodies* which from time to time visit the earth, sweeping through the sky with immense velocity in every direction, and remaining visible but a few moments; they are generally attended by a luminous train, and during their progress explosions usually occur, followed by the fall of stones, to which the name of *aërolites* is given, from the Greek words *aër*, atmosphere, and *lithos*, a stone.

503. FACTS. At noon, on the 7th of Nov. 1462, at Ensisheim, in Germany, a loud explosion was heard in the air, and a *stone seen* to fall which buried itself deep in the earth. It weighed 260 lbs., and by the order of the Emperor Maximilian, was suspended in the church at Ensisheim, where it remained until the French revolution. A portion of it is now in the Parisian museum, and another in the Imperial Cabinet at Vienna.

On the 21st of June, 1635, a *fiery mass* was seen passing over the Veronese territory with such velocity, that the eye could scarcely follow its motions. Loud

---

What is the subject of part sixth?
What of chapter first?
Define meteorites.
What are aërolites?
Relate the account of the meteorite of Ensisheim.
Of that of Verona.

explosions were heard, and a *large stone fell* near the Benedictine Convent, about six miles from Verona.

504. At half past six o'clock, on the morning of the 14th of Dec. 1807, a meteorite was seen rushing from north to south, over Weston, in the State of Connecticut; its *apparent diameter* being equal to *one-half*, or *two-thirds*, that of the *full moon.* As it passed behind the clouds, it appeared like the sun through a mist, and shone with a mild and subdued light; but when it shot across the intervals of clear sky, the glowing body flashed and sparkled like a firebrand carried against the wind. Behind it streamed a *pale, luminous train*, tapering in form, and *ten or twelve times as long as its diameter.*

The meteorite was visible for the space of half a minute, and just as it vanished gave *three, distinct bounds.* About thirty seconds after its disappearance, *three heavy explosions* were heard like the reports of a cannon, succeeded by a *loud, whizzing noise.* Directly after the explosions, a person of the name of Prince heard a sound resembling that occasioned by the fall of a heavy body, and upon going from the house perceived a fresh hole in the turf, at the distance of twenty-five feet from the door. At the bottom of the hole, two feet below the surface, an *aërolite* was discovered which weighed nearly 35 pounds.

Another mass, which was dashed to pieces upon a rock, was judged, from the fragments collected, to have weighed *two hundred pounds.* Other aërolites fell in various parts of the town. The stones, at the time of their descent, were *hot* and *crumbling,* but gradually hardened upon exposure to the air.

505. At Futtypore, in India, on the 5th of Nov. 1814, a meteorite was seen, shortly after sunset, shooting swiftly towards the north-west. It appeared as a *blaze of light* surrounding a *red globe* of the *apparent size of the moon.* As it proceeded on its course, loud explosions were heard, resembling the sound of distant artil-

---

Of that of Weston.
Of that of Futtypore.

lery, and a *stone fell*, which, in its descent, emitted sparks like those proceeding from a blacksmith's forge. When first discovered, the *aërolite* was *hot*, and exhaled a strong sulphurous smell.

On the 11th of Dec. 1836, just before midnight, a meteorite of extraordinary size and brilliancy was seen over the village of Macao, in Brazil, traversing a cloudless sky. It burst with a sharp, loud noise, and a *shower of stones fell* within a circle of *ten leagues*. The aërolites varied in weight from *one pound* to *eighty*, and descended with such force as to break through the roofs of houses, and bury themselves deep in the sand.

506. Dr. Chaldni has compiled an extensive catalogue of instances when meteorites have been *seen;* from which it appears, that these extraordinary bodies have been noticed from the *earliest ages*, and in all *parts of the world;* and, since attention has been drawn to the subject, scarcely a year now passes without one or more well-attested cases of the fall of aërolites.

507. SIZE OF METEORITES. We must not confound the magnitude of the *meteorite* with that of the *aërolite*, for the latter is nothing more than a fragment thrown off from the former and falling to the earth, while the main body sweeps onward in its course.

The diameter of the Weston meteorite was computed to be 300 feet, and that of the meteorite observed by Mr. Cavallo, at Windsor, on the 18th of Aug. 1783, was calculated by this gentleman to be no less than 3210 feet, or more than *three-fifths of a mile*. Mrs. Somerville mentions one that was estimated to weigh nearly 600,000 tons.

508. ALTITUDE. The height of meteorites above the earth has been estimated, and found to vary from 18 to 70 or 80 miles. According to the calculations of Dr. Bowditch, the meteorite of Weston approached within 18 miles of our globe, and one mentioned by Mrs. Somer

---

Of that of Macao.
When and where have meteorites appeared?
What is said regarding their size, altitude and velocity?

ville, came within the distance of 25 miles. An instance is given by Dr. Halley of a meteorite that exploded at an elevation of 69 miles, with a report like that of a cannon.

509. VELOCITY. The velocity of these bodies is generally somewhat more than 300 miles per minute, though many cases have occurred of far greater speed; the meteorite just mentioned, that came within 25 miles of our earth, moved at the rate of 1200 miles per minute.

510. If a body in the atmosphere is seen at the *same time* by two observers upon the earth at *different stations*, and its angular elevation taken at *both stations;* its height in miles and feet is easily ascertained by the aid of trigonometry, when the *distance between the two stations* is known. If the body is in motion, and its position noted at the *moment* of its *appearance* and *disappearance*, the distance it travels in this interval, or the *length of its visible path* can be obtained, when its height has first been computed. The *speed* is ascertained by dividing the length of the visible path by the number of seconds during which the body is seen. The *magnitude* is easily obtained by trigonometrical calculations, when the *distance* of the body and its angular diameter is known. In this manner computations are made upon *meteorites, shooting-stars* and the *aurora borealis.*

511. From their sudden appearance and extreme velocity, all observations upon these phenomena are liable to great inaccuracy, and estimates of magnitude, velocity and height, derived from such observations must be received with much allowance, and are to be regarded only as approximations, more or less near to the truth.

AEROLITES.

512. FORM. The greater number of aërolites, according to Schreibers, have always the *same general form*, which is that of an *oblique* or *slanting pyramid.*

In what manner are calculations made respecting the size, altitude, speed, &c., of bodies high above the earth?
What is said as to estimates of this kind?
What is the form and external appearance of aërolites?

They are also alike in *external appearance*, presenting to view a *black, shining crust*, as if the body had been coated with pitch. This crust is not greater than the two-hundredth part of an inch in thickness, its composition is *identical* with that of the *mass*, it bears the marks of fusion, and strikes fire with the flint. When *broken*, the surface of the fracture displays the color of an *ash-grey*.

"Distinct aërolites," says Berzelius, the celebrated chemist, "are frequently so like one another in color and external appearance, that we might believe them to have been struck out of one piece."

513. COMPOSITION. According to Berzelius, aërolites consist of eighteen elementary substances. A nineteenth has since been discovered, and perhaps *two* others. They are remarkable for containing *malleable metallic iron, nickel*, and *chrome*. Their specific gravity varies from 3.35 to 4.28.

514. These *common characteristics* indicate a *common origin*, and this we are led to seek beyond the earth, inasmuch as the composition of aërolites is totally different from that of any stony mass, forming a part of the crust of the globe. *Malleable metallic iron* is *rarely*, if *ever*, found in *terrestrial* substances, *nickel* is extremely scarce, and has *never been discovered* on the surface of the earth; and *chrome* is, if possible, *still more rare*.

515. It sometimes happens, though seldom, that the aërolite consists *almost entirely of metallic iron*. On the 26th of May, 1751, a meteorite burst with a tremendous report, over Hradschina, in the district of Agram, in Upper Sclavonia. *Two* fragments were seen rushing to the earth, the largest of which struck deep into the soil. This mass weighed 71 lbs., exhibited evident traces of fire, and, upon being analyzed, gave, out of every 100 parts, 95.5 of *iron* and 3.5 of *nickel*. A portion of the

---

What is their composition?
What do these common characteristics indicate?
Where must it be sought—and why?
Of what does the aërolite sometimes consist?
Give the instance.

iron being finely polished, and corroded with acids, a most beautiful *crystaline structure* was revealed, branching in every direction over the surface. This peculiarity belongs to most of the specimens of the iron of meteorites. (See Fig. 41.)

Fig. 41.

CRYSTALINE STRUCTURE OF THE METEORIC IRON OF TEXAS.
(*Copy of an Impression taken from the Iron.*)

516. METEORIC IRON. From the peculiar constitution and structure of aërolites, we are enabled to detect the *meteoric origin of masses of iron* which are occasionally found scattered over the surface of the earth, in all quarters of the globe. For since they possess the *same elements* as the iron of aërolites, *combined* in the *same* manner, and as no such masses have ever been taken from mines, we must necessarily conclude, that they were once exploded from a meteorite, though no record exists of their fall.

517. Humboldt relates, that in Mexico, near the envi-

What structure did this iron possess?
Why are certain masses of iron supposed to have a meteoric origin?

rons of Durango, is an *enormous mass of malleable iron and nickel,* which possesses *exactly* the *same composition* as the fragment that fell at Agram.

A mass of *metallic* iron, weighing 1544 lbs., was discovered by Prof. Pallas, in 1771, at Krasnojark in Siberia. It was regarded by the Tartars as a sacred object, and according to their traditions had fallen from heaven.

The famous mass of *malleable* iron which was found in Texas in 1808, and is now in the cabinet of Yale College, weighs 1635 lbs. It contains *nickel.*

518. During an expedition in South Africa, Sir James Alexander discovered, near the Great Fish river, a considerable tract of country, over which fragments of *metallic iron* were scattered in profusion; a specimen analyzed by Sir John Herschel, was found to possess *nickel,* thus proving conclusively the meteoric origin of the masses.

519. ORIGIN OF METEORITES. Natural philosophers have advanced *five* hypotheses, to account for the origin of these extraordinary bodies.

* 1st. *That they are ejected from terrestrial volcanoes.*

2d. *That they are produced in the atmosphere, being formed from the gases exhaled from the earth.*

3d. *That they are thrown from lunar volcanoes.*

4th. *That they are terrestrial comets revolving about the earth like the moon.*

5th. *That they are celestial bodies revolving about the sun like the planets, and encountered by the earth in its annual progress.*

520. FIRST HYPOTHESIS. The first supposition cannot be maintained, since it is impossible for the volcanoes of the globe to hurl to the height of *twenty miles* masses of the size of meteorites; besides, the composi-

---

Illustrate from the several instances given.

How many hypotheses have been advanced, to account for the origin of meteorites, and what are they?

What are the objections to the first hypothesis?

tion of the latter is *entirely different* from all *volcanic products*.

521. SECOND HYPOTHESIS. The second is likewise untenable. *Nickel*, according to high chemical authorities, has never been raised in vapor; even under the intense heat of volcanoes. A mass of matter formed in the air, must therefore be destitute of nickel, an element which meteorites invariably possess. Moreover, such a body, in its descent, would fall *perpendicularly* to the ground by the action of gravity, and not sweep along, as did the Weston meteorite, in a direction *nearly parallel* to the surface of the earth.

522. THIRD HYPOTHESIS. In regard to the third hypothesis, it has been shown, by calculations of La Place, and other eminent mathematicians, that a mass projected from the moon, with a velocity of 10,660 *feet per second*, would pass beyond the point of the moon's attraction, and either fall to the globe in the space of two days and a half by the force of gravity, or revolve about the earth like the moon. It is not therefore *impossible*, that such an event might *occasionally* occur, but it is *utterly improbable* that meteorites originate in this manner.

523. Omitting other objections to this hypothesis, the *size* and *number* of meteorites constitute an insuperable difficulty. It requires a strong faith to believe, that such masses, as have been described, could be hurled from a lunar volcano, at the rate of not less that 10,000 feet per second; a speed *five times* greater than the *highest velocity* of a cannon-ball.

524. The *number* of meteorites must be also very great; for they have been seen from the earliest ages and in all inhabited quarters of the globe occasionally traversing the heavens, and those which have been noticed are probably only a part of the *actual number* that have visited the earth. Many must have passed unseen over the broad expanse of ocean, or crossed vas'

What to the second?
What argument is advanced in favor of the third hypothesis?
What arguments against it?

tracts of uninhabited lands, leaving no trace of their existence, except those masses of meteoric iron, which from time to time are brought to light. If therefore the lunar theory is adopted, we can scarcely avoid the conclusion, that the moon has been ejecting for ages, so *many*, and *such vast* masses of matter, as must have sensibly diminished her bulk, and occasioned derangements in her system—results at *variance with all observations*.

525. FOURTH HYPOTHESIS. The fourth hypothesis, which is that of President Clap, of Yale College, affords a more reasonable explanation of the phenomena of these extraordinary bodies than any of the preceding. Under this view the earth is supposed to possess a system of comets like the sun. The solar comets revolve about their primary in very extended orbits; at one part of the course approaching so near the sun as almost to strike its surface, and during the remainder, sweeping far out of sight beyond the path of the planets, continuing invisible for *years* and even *ages*. In like manner *meteorites* are supposed to *revolve about the earth;* their *size* and *periods of revolution* being proportioned to the smallness of their primary. Moving also in very elliptical or oval orbits, they are too distant to be visible during the greater part of their course, but at *one point* of their path approach very close to the earth, and enter its atmosphere.

On account of the immense velocity of the meteorite, the air is imagined to be condensed before it to such a degree, that heat is evolved of sufficient intensity to inflame the mass at its surface, while during this combustion gases are generated, which by their expansive energy, produce *explosions*. By the strength of this disruptive force, glowing fragments are detached from the surface and fall to the ground, while the meteorite itself passes onward on its course.

526. It has been calculated, that the velocity of a

Which hypothesis affords a more reasonable explanation?
Explain it fully.

body revolving about the earth must not be *less* than 300 miles per minute, nor *greater* than 420. Were it *less* than 300 miles, the mass would *fall* to the earth by the action of gravity; and if the rate exceeded 420 miles, it would pass away from the globe and *never return*. Within these limits, allowance being made for the motion of the earth in its orbit, and the resistance of the air, the body would revolve around the earth like the moon, approaching very near to its surface at stated periods.

527. In support of this hypothesis it is urged, that the velocity of meteorites, in general, is somewhat more than 300 miles per minute, though doubtless cases have occurred in which their speed was far greater.

The combustion of the meteorite, through the agency of a condensed atmosphere, is by no means improbable; for though the medium in which it moves is *exceedingly rarefied*, yet the velocity of the body is amazing; and it can easily be shown by calculation, that from the condensation thus effected, an intensity of heat would be developed of which we have no conception. (Art. 551.) Moreover, as *silica*, *magnesia*, and *potassa* are found in meteorites, it has been conjectured, that they may originally exist there in the state of *pure metals;* and, that when the meteorite enters our atmosphere, combustion arises from the extraordinary affinity of these substances for *oxygen*.

In those instances where meteorites move at a greater rate than 420 miles per minute, they are supposed either to revolve about the sun, and that the earth occasionally meets them in her annual progress; or to wander through space, until they come within the superior attraction of some other orb, and are then compelled to revolve around it.

What calculation has been made in respect to a body revolving about the earth?

What facts and suggestions are adduced in support of Pres. Clap's hypothesis?

What is said of meteorites moving at a greater rate than 420 miles per minute?

528. FIFTH HYPOTHESIS. The last hypothesis is that of Chaldni, and is explained in Art. 555. In this the *ignition* and *explosion* of the meteorite are attributed to precisely the same causes as those assigned in the fourth hypothesis.

CHAPTER II.

OF SHOOTING-STARS AND METEORIC SHOWERS.

529. Shooting-stars or *meteors* differ from meteorites in several particulars. They commonly possess a *superior* velocity, and their *altitude* is generally *greater:* bursting from the clear sky, they dart along the heaven like a rocket, *consuming themselves in their course*, and leaving behind a luminous train, which gradually vanishes in a short time. Unlike the meteorite they usually pass away *without any explosion, and no portion of the body ever reaches the earth*. Besides, they are far more *numerous* and *frequent;* appearing almost every night, and at times descending in such multitudes that the heavens are illumined *for hours* with their glowing trains.

530. ALTITUDE. In order to investigate the phenomena of shooting-stars, Brandes and Benzenberg, two German philosophers, made a series of simultaneous observations in the fall of the year 1798. On six evenings, between September and November, 402 shooting-stars were beheld, and of these *twenty-two* were so identified, that their *altitudes*, at the moment of their extinction, could be readily computed. They were found to be as follows:

7 disappeared at altitudes *under* 45 miles.
9   "   "   *between* 45 and 90 miles.
6   "   "   *above* 90 miles.

Of what does chapter second treat?
In what particulars do shooting-stars and meteors differ from meteorites?
Relate the account of the observations of Brandes and Benzenberg, for determining the altitudes of shooting-stars?
Give their results.

The least and greatest elevations were *six miles* and *one hundred and forty.*

531. In 1823, the investigation was renewed by Brandes, at Breslau and the neighboring towns, on a more extended scale. Between April and October, 1800 shooting-stars were seen at the different stations. Out of this number, 98 were observed *simultaneously* at more than one station, and afforded the means of estimating their respective *altitudes.* The results were as follows:

4 disappeared at altitudes *under* 15 miles.
15 " " *between* 15 and 30 miles.
22 " " " 30 " 45 "
33 " " " 45 " 70 "
13 " " " 70 " 90 "
11 " " *above* 90 "

Out of the last eleven, *two* vanished at an elevation of 140 miles, a *third* at 220 miles, a *fourth* at 280 miles, and a *fifth* at 460 miles.

The height of *four shooting-stars* noticed by Professors Loomis and Twining, in December, 1834, varied from 54 miles to 94.

532. Similar observations were made in Switzerland, on the 10th of August, 1838, by Wartman and others. A part of the observers stationed themselves at Geneva, and the rest at Planchettes, a village about sixty miles to the north-east of that city. In the space of seven hours and a half, 381 shooting-stars were seen at Geneva, and in five hours and a half 104 at Planchettes. All the circumstances attending their appearance were carefully noted, and their average height was computed at *five hundred and fifty miles.*

533. VELOCITY. In the *first* series of observations made by Brandes and Benzenberg, only *two* shooting-stars afforded the means of determining their *speed;* one possessed a velocity of 1500 *miles per minute,* and

---

Give those of Loomis and Twining.
Give those of Wartman, at Geneva.
What is their velocity according to the observations of Brandes and Benzenberg, and Quetelet?

that of the other was between 1020 and 1260 miles *per minute.*

In the *second* series, undertaken in 1823, the estimated rate of motion varied between 1080 and 2160 miles *per minute.* At Belgium, in 1824, M. Quetelet obtained observations upon six of these singular bodies, from which he was enabled to deduce their respective velocities, which were found to range from 600 to 1500 miles *per minute.*

534. COURSE. Of thirty-six stars, whose paths were ascertained by Brandes, the motion in twenty-six cases was *downward,* in one *horizontal,* and in the remaining nine, *more or less upward ;* nor did they always move in *straight lines ;* for the paths of some were *curved,* either upwards or sideways; while others proceeded in a *serpentine* course. Their general direction was from *north-east* to *south-west.* Several examples have been given by Chaldni, where the luminous body described a *semicircle,* first *rising* and then *falling.*

535. MAGNITUDE. The size of shooting-stars is variable. Fire-balls, which are regarded as nothing more than large meteors, have sometimes appeared of a magnitude almost incredible. During the remarkable shower of meteors, on the 12th and 13th of November, 1833, luminous globes, apparently as large as Jupiter and Venus, were seen darting through the air in all directions. About three o'clock on the morning of the 13th, a splendid body which appeared equal in size to the full moon, swept across the heaven from east to west. If the distance of this meteor was only *eleven* miles, its diameter must have been 528 feet, or *one tenth of a mile.* Amid the shower of stars that occurred in 1799, meteors were observed by Humboldt, apparently *twice the size of the moon.*

536. On the evening of the 18th of May, 1839, a meteor of extraordinary magnitude passed over the

---

What is said respecting the course of shooting-stars?
What of their magnitude?
Relate the account of the meteor of the 18th of May.

Northern States and a part of Canada. From the facts which he collected, Prof. Loomis estimated its diameter at 1320 yards, or *three quarters of a mile*. Its velocity was computed by this gentleman to be nearly 2100 miles a minute, its height to be 30 miles, and the length of its path 200 miles. The meteor was followed by a train of inconsiderable extent, probably formed of the detached portions of the body which fell behind.

537. SPLENDOR. At times these luminous bodies present a spectacle of surpassing beauty, from their *brilliant coruscations, extended trains,* and *rich diversity of colors*. During the month of April, 1832, a globular ball of fire, apparently a *foot in diameter*, passed over Torhut, in India, early in the morning. Behind it streamed a train of dazzling light, which appeared to be several yards in length. The meteor illumined the surrounding country to a great distance, and after remaining visible for the space of five seconds, exploded *without noise*, like a rocket, throwing out numerous coruscations of intense splendor.

In May of the same year, and at the same place, a similar body was seen moving rapidly through the air, from north to south. It glowed with a brilliant mixture of *green* and *blue light*, and vanished in about three seconds, leaving a luminous train of great length.

538. During the nights of the 9th and 10th of August, 1839, many shooting-stars of singular beauty were seen by Mr. E. C. Herrick, of New Haven. One flashed with a *golden green* light, and another sparkled with *green* and *blue*. Meteors *entirely green* have at times been noticed. A meteor which swept over Kensington, near London, in 1839, as brilliant as Jupiter and apparently of greater size, presented the rare combination of white light in the mass, with one edge *red* and the opposite of a *deep blue* or *purple*.

On the morning of the 13th of November, 1833, a most brilliant meteor was seen by Prof. Twining, de-

What is said of their splendor?

scending towards the earth with majestic rapidity. Its apparent *size* was one-fifth that of the moon, and its *color* a *deep red.* It vanished when near the ground, leaving behind a fiery train of the same hue, excepting that it displayed the *prismatic tints,* especially at the point where the meteor expired.

539. The usual color of meteors is that of a *phosphoric white* tinged with *red.* The trains generally vanish in a few *seconds,* but they have been known to last for the space of *seven* minutes, and even *fifteen.* Their light (as we have just seen) is not invariably of one hue, for at times it presents to the eye all the rich tints of the rainbow.

METEORIC SHOWERS.

540. The wondrous display of meteors in 1833, drew the attention of philosophers to the subject of shooting stars, and, from the results of subsequent researches and observations, there is now reason to believe, that certain epochs exist when these luminous bodies appear in greater numbers than usual, and that sometimes at the return of these periods they *literally* descend to the earth in *showers.* The best ascertained periods are those of the 12th and 13th of November, and the 9th and 10th of August.

541. NOVEMBER EPOCH. On the morning of the 12th of November, 1799, an extraordinary display of this nature was seen by Humboldt and Bonpland, at Cumana, in South America. During the space of *four hours* the sky was illumined with *thousands* of shooting-stars, mingled with meteors of vast magnitude. This phenomenon was not confined to Cumana, but extended from Brazil to Greenland, and as far *east* as Weimar, in Germany.

On the 13th of the same month, in 1831, a meteoric shower occurred at Ohio, and also near Carthagena, off the coast of Spain. At the latter place, luminous meteors of large size were beheld, one of which left behind it an

---

Are meteors at all times equally abundant?
What two great epochs exist?

enormous train, tinted with prismatic hues, its trace continuing visible for the space of *six minutes*. On the same day of the following year, vast numbers of shooting stars fell at Mocha on the Red Sea, upon the Atlantic ocean, and in Switzerland. The same brilliant spectacle then appeared in various parts of England; the sky being illumined soon after midnight by the rushing of thousands of meteors in every direction.

542. But by far the most magnificent display of this kind occurred on the night of the 12th and morning of the 13th of November, 1833. It extended from the northern lakes to the south of Jamaica, and from 61° W. Long. in the Atlantic to about 150° W. Long. on the Pacific ocean near the equator. For the space of *seven hours*, from 9 P. M. to 4 A. M., the heavens blazed with an incessant discharge of fiery meteors, that burst in countless numbers from the cloudless sky. At times they appeared as thick as snow-flakes falling through the air, as large and as brilliant as the stars themselves; and it required no vivid imagination to suppose, that these celestial bodies were then actually rushing towards he earth.

543. VARIETIES. The luminous bodies of this shower seemed to be divided into *three* kinds. The first consisted of *bright lines* traced through the sky, as if by a point. The second of *fiery balls*, that occasionally darted across the heavens, trailing behind them extended and luminous trains, which generally continued visible for many minutes. The third of *radiant bodies*, that *continued almost immovable* for a considerable time.

544. Meteors of the *first class* occurred in great abundance. At Union Town, Pennsylvania, they were seen shooting along *like streams of fire* with the rapidity of lightning; often crossing half the visible heavens in less than a second.

At New York, about a quarter past five o'clock, a meteor of the *second class* was beheld rushing from the

---

Describe the meteoric showers of November, 1799, 1831, 1832 and 1833. In the shower of 1833 how many kinds of meteors were noticed? Describe them, and give the instances.

zenith, and marking its course by a *fiery line* apparently two or three inches wide. After passing downward to a considerable distance, it formed into a *ball* of the *apparent size of a man's hat*, and then returning on its path, assumed a *serpentine figure*. It lay extended through the sky for the space of *several minutes*, and then struck off to the west.

A meteor of the *third* kind was visible in the northeast, at Poland, Ohio, for *more than an hour*. It first appeared in the form *of a pruning hook*, apparently twenty feet long, and eighteen inches broad, and shone with great splendor. At Niagara Falls, at two o'clock in the morning, an extended *luminous body like a square table* was noticed in the zenith. It remained for a time nearly stationary, sending out on every side broad streams of light.

545. It was distinctly noticed by many attentive and accurate observers, that all the meteors appeared to emanate from a certain region, situated in the constellation Leo; and that during the whole display this point was *stationary among the stars* for more than *two hours;* thus proving, that the source of the meteoric shower was *beyond the atmosphere of the earth;* for had it been *within*, it must have moved *eastward*, in the direction of the earth's daily motion.

546. For *four successive* years, after the great November shower of 1833, an unusual number of meteors was observed in America at this period. The phenomenon ceased, upon this continent, in 1838; but an extraordinary display then occurred at Vienna, more than a *thousand* meteors falling in the course of *six hours*.

547. AUGUST EPOCH. The *second* meteoric period occurs on the 9th and 10th of August. It was first distinctly announced in 1827 by Thomas Foster of London,

---

What fact was distinctly noticed by attentive observers?
Where is this point situated?
What is inferred from the circumstance that it was stationary?
For how many years after 1833 did this phenomenon appear?
When does the second meteoric period occur?
By whom was it first announced?

in his Encyclopedia of Natural Phenomena. The number of meteors observed at this epoch is probably five or six times more than the usual nightly average, which has been estimated by Mr. E. C. Herrick, of New Haven, at not more than *thirty per hour for four observers.*

548. From 1836 to the present year, scarcely a season has passed without an unusual display of meteors at this period, in some quarter of the globe.

On the 9th of August, 1839, four observers at New Haven beheld 691 shooting-stars in the course of five hours, a *third* part surpassing in brightness stars of the *first magnitude.* On the ensuing night, 491 were seen in the space of *three hours,* by the same number of observers; and at Vienna in Austria, during the same evening, shooting-stars descended at the rate of *sixty per hour.*

Upon the annual return in 1842, 490 meteors fell at Parma in Italy, and 779 at Vienna. Many were likewise seen at Brussels. At New Haven, in the space of fifty minutes, 89 were seen, one of which equaled Jupiter in splendor.

In 1847, at Manlius, N. Y., 415 meteors were seen on the morning of the 11th of August in the course of *two hours,* commencing at midnight and ending at 2 o'clock A. M. On the 10th of August, 1848, 475 meteors were noted at New Haven, in the space of *two hours and a half,* by Mr. E. C. Herrick and three other observers. Many of them exceeded in brilliancy stars of the first magnitude. In France, on the same night, 414 shooting-stars were beheld by two observers, within a period of *three hours and a quarter.*

549. Like the meteors of November, those of August appear also to radiate from a *small space* in the heavens, which has been referred, by all observers, to the constellation Perseus.

Shooting-stars have likewise been found to be more

What is said in regard to the recurrence of this shower?
State facts.
What is said respecting the source of the August meteors?
Where is it situated?

than usually abundant on the 18th of October, the 6th and 7th of December, the 2d of January, the 20th of April, and from the 15th to the 20th of June.

550. ORIGIN. Prof. Olmsted, who was the first to present his views upon the extraordinary phenomenon, which occurred on the 12th of November, 1833, has arrived at the following conclusions from a very extensive examination of facts.

That the source of the meteors is a *body* possibly of great extent, composed of matter exceedingly *rare* like the tail of a comet. That it revolves about the sun within the orbit of the earth, its period of revolution being probably a little less time than a year,

That in consequence of its proximity on the night in question, the *extreme parts* of the body were detached and drawn towards our globe, by the *force of gravity*.

That its altitude above the surface of the earth, at its nearest point, was about 2238 miles; and that the descending fragments entered the atmosphere with a velocity ranging from about *fourteen* to *twenty miles per second*.

That these *fragments were combustible*, and in consequence of their amazing velocity, the air was so powerfully compressed before them, that they *took fire*, and were consumed before reaching the earth.

551. This last conclusion will appear by no means incredible, when the following considerations are taken into view.

By suddenly forcing down a solid piston to the bottom of a cylinder, in which it moves air-tight, sufficient heat can be evolved to ignite tinder; and this occurs, when the air within the cylinder is compressed to *one-fifth* of its original volume. Upon the supposition, that the descending fragments compressed the rarefied atmosphere at the height of 35 miles only to the *density of common air*, the amount of heat developed would be 46,080°

What is said of other periods?
Detail Prof. Olmsted's theory.
What is said respecting the amount of heat developed by the condensation of the atmosphere?

Fah.; an intensity nearly *three times greater* than the highest temperature of a glass-house furnace, which is 16,000° Fah.

552. If the *nebulous* body revolves about the sun in a period somewhat less than a year, it tends to explain the occurrence of shooting-stars at *all seasons* (since the earth and the nebulous body would then be always comparatively near each other), and will also favor the explanation of the meteoric showers which have happened towards the end of April.

553. Prof. Olmsted has been led to suppose, from the whole course of his observations, that the *nebulous body* in which the meteors originated, might be identical with the *zodiacal light*. In a late article published by M. Biot, this distinguished philosopher also maintains, that meteoric showers are occasioned by the zodiacal light coming in periodic contact with the atmosphere of the earth.

It is not regarded by Prof. Olmsted as essential to the truth of his theory, that a shower of meteors should occur upon the 13th of *every* November

554. In order to account for shooting-stars in general, including alike their *ordinary* and *extraordinary* displays, and embracing the several epochs, the views of Chaldni have been adopted by Arago and other eminent philosophers

555. CHALDNI'S THEORY. This theory consists in supposing, that, besides the planets, *millions of small bodies* are constantly revolving about the sun, which become ignited when they enter the terrestrial atmosphere. They are not considered to be *uniformly* spread throughout space; but in some regions to be *diffusely scattered*, and in others *grouped* together in vast multitudes, forming *zones or rings around the sun;* many of which cross the path of the earth.

The ordinary, nightly phenomenon of shooting-stars,

If the nebulous body revolves about the sun in a little less time than a year, what does it tend to explain?
What is M. Biot's opinion? What is Chaldni's theory?

is then imagined to arise, when the earth, in her progress through the heavens, traverses those regions which contain only a *few* of these *bodies;* but when the *zones* are encountered, and the globe passes amid countless numbers, the display is proportionally greater, and the meteors occasionally descend in magnificent showers. Amid this vast collection solid masses of considerable size are supposed to exist, and should one of these enter the atmosphere of the earth, a *meteorite* with all its splendors sweeps across the sky.

Such at present is the general state of our knowledge in regard to shooting-stars.

CHAPTER III.

OF THE AURORA BOREALIS OR NORTHERN LIGHT.

556. The Aurora Borealis is a luminous appearance in the northern sky, which presents, when in full display, a spectacle of surpassing splendor and beauty. It has in all ages been an object of wonder and mystery, and still continues so; for although many valuable facts have been brought to light by the investigations of science, the cause of this brilliant phenomenon is yet involved in obscurity.

557. Constitution. Notwithstanding its fantastic motions, and momentary changes in brightness and color, the aurora, according to the best observations, still preserves, amid all its fluctuations, certain invariable characteristics of form and position. It consists of a *dark segment,* an *arch of light, luminous streamers,* and a *corona* or *crown.*

558. Dark Segment. All observers in the high latitudes of Europe, agree in stating, that before the

---

What does chapter third treat of?
What is the Aurora Borealis?
Of what does it consist?
Describe the dark segment.

aurora appears, the sky in the northern horizon assumes a *darkish hue*, which gradually deepens, until a *circular segment* is formed, bordered by an arch of light, extending from east to west. The segment presents the appearance of a cloud, its tint is *light* in the lower latitudes, and grows darker as we advance to the north, up to a certain limit; after this the reverse occurs, and when high latitudes are attained it becomes so faint as to be scarcely visible. At Upsal and Christiana it is sometimes *black* or of a *deep gray*, which changes into a *violet*.

During a splendid aurora, that occurred at Toronto in Dec. 1835, and which is described by Capt. Bonnycastle, a *dark, black changing mass*, was visible below the luminous arch, (fig. 42,) and in a remarkable phase of the aurora, when several bright bows were seen at once, the interval between the *second* and *third* assumed a *blackness* of the *deepest intensity*.

AURORA SEEN AT TORONTO.

559. A difference of opinion exists in regard to the nature of this segment. From numerous observations made at Dorpat in Russia, Struve infers, that the dark-

Is it *real* or *imaginary*?

ness is simply the effect of *contrast* with the luminous arch; while, from equally extensive researches at Abo in Finland, Argelander concludes, that the segment is something *real ;* since the portion of the sky it occupies, is darker than common, *before* the bright bow of the aurora appears.

560. ARCH OF LIGHT. The dark segment is bounded by a *luminous arch or bow*, varying in width from *one* to *three apparent diameters* of the *moon*. Its lower edge is clearly defined, but the upper is only so when the arch is narrow, for as the width increases, it gradually blends with the brightness of the sky. The color of the bow is a pale white, which becomes more pure and brilliant near the polar regions.

According to the most accurate observations, this arch has a tendency to place itself at *right angles* to the *magnetic meridian,* or in other words, to the direction of a compass-needle at rest. (C. 985.) This fact was particularly noticed by Lieutenant Hood, who accompanied Franklin in his northern expedition in 1819.

561. The centre of the auroral arch probably coincides with the north magnetic pole of the earth, which is situated in 70° N. Lat. In our own country, the compass-needle points to the north, and the arch crosses the heavens from *east* to *west ;* but in some parts of Greenland, the needle is directed to the *west,* and the arch is then seen extending from *north* to *south.*

In the year 1838, when Simpson wintered at Fort Confidence, in 66° 54′ N. Lat., he found the needle always pointing to the *north-east,* and the auroral arches invariably spanning the heavens at *right angles,* from *north-west* to *south-east.*

At Melville Isle, in 74° 30′ N. Lat., the luminous arches were seen by Parry in the *south ;* the north magnetic pole of the earth being then in that direction.

562. This beautiful bow of light is not stationary,

Describe the arch of light. Its color and position.
What is its position in some parts of Greenland?
What was its position at Fort Confidence and at Melville Isle?
Is the arch of light stationary?

but frequently rises and falls; and when the aurora appears in great splendor, several arches are seen at the same time crossing the sky, ascending gradually from the horizon to the zenith, and passing over in succession with their summits moving in or parallel to the magnetic meridian; presenting to the eye broad belts of light, increasing in brightness as they approach the zenith.

563. No less than *five* such arches were seen at *once* by Lieut. Hood; but similar phenomena, of far greater beauty, were witnessed by M. Lottin at Bossekop, in West Finmark, during the winter of 1838-9. (Figs. 43, 44.)

Fig. 43.

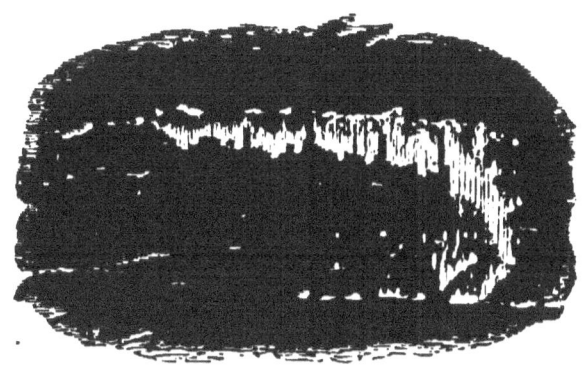

AURORA SEEN AT BOSSEKOP.

Fig. 44.

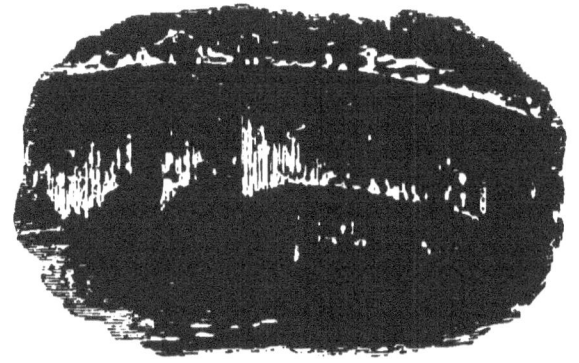

AURORA SEEN AT BOSSEKOP.

What phenomena were beheld by Lieut. Hood and M. Lottin?

On one occasion, as many as *nine auroral arches* were visible, separated by distinct intervals, and in their arrangement resembling magnificent curtains of light, hung one behind and below the other, their dazzling folds extending completely across the sky.

564. STREAMERS. Although the luminous arch preserves, in the main, its curved form, it is subject to constant changes. Now at one extremity, now at the other, and again at intermediate points, a cloud of light will break suddenly forth, separating into *rays* which stream upward like tongues of fire, moving at the same time backwards and forwards, along the auroral bow.

The origin of the *streamers* is in the *luminous arch*, from which they rise in the form of tapering rays or pencils of light, ever in motion, and continually varying in brilliancy, number, magnitude, and color. At one moment, a ray is just visible above the arch, faintly glowing in the sky; at the next it is seen shooting upward in a pyramid of flame and at the same time moving majestically across the heavens. As suddenly its brightness fades, and as quickly it is again beheld, flashing forth with renewed splendor.

565. COLOR. During the extraordinary displays of the aurora in our own latitude, the sky is frequently seen suffused with a flush of *rosy light*, while the *streamers* assume a *crimson hue*. In that which occurred on the night of the 14th of November. 1837, the upper extremities of the streamers were of the deepest *scarlet*, while below they were brilliantly *white*. But the richest tints appear in the arctic regions. In the auroras witnessed at Bossekop, the rays, at their base, glowed with a *blood-red hue*, the middle was of an *emerald green*, and the rest of a *pure transparent yellow*. During a brilliant display that occurred at Fort Confidence, on the 5th of March, 1839, the rays were tinged with *red, purple*, and *green*.

566. CORONA OR CROWN. The vivid rays that dart

What is said in regard to the streamers, their origin and color?

forth from the luminous arch not unfrequently unite at a point near the zenith; forming a brilliant mass of light which is called the *corona* or *crown*. The aurora then appears in its greatest splendor; the sky resembles a fiery dome, and over the streamers, which seem like pillars of variegated flame supporting the corona, radiant waves and flashes of light pass in quick succession. The luminous columns at this time are apparently shaken and wave with a *tremulous* motion; whence they have received, under these circumstances, the name of *merry dancers*.

At Bossekop, this radiant wave was seen by Lottin, crossing and re-crossing with rapid undulations the whole broad field of auroral light. These coruscations are generally attended with color.

567. When, in the northern hemisphere, a needle is delicately balanced upon a horizontal axis, its north end immediately *dips downward* upon its being *magnetized*. Such an instrument is called the *dipping-needle*. (C. 998.) The *streamers* of the aurora assume the *same direction as the dipping-needle*, and are *parallel* to each other; hence the corona is not formed by any *actual union* of the streamers near the zenith. It arises from an *optical illusion*. When we look across an extensive field of corn, the rows, at *their remote ends*, seem to approach each other, as if converging to a point; though we know that they are three or four feet apart, throughout their whole distance. In like manner when we gaze at the auroral *streamers* with their bases at the horizon and their summits at the zenith, they will in like manner apparently converge to one point, forming the *corona*, whose centre is in the line of the dipping-needle.

568. Within the *dark segment* streamers of the same color are frequently seen, rising and falling like columns of smoke, changing their hue in a moment, and possessing all the motions of the luminous rays. Like the

---

Describe the corona.
When does it appear?
How is it formed?
What is observed within the dark segment?

latter, their line of direction is parallel to that of the dipping-needle.

569. EXTENT. The aurora is not a local appearance; for it is beheld simultaneously in places widely separated from each other. Thus, on the 5th of January, 1769, the same aurora was seen in France and Pennsylvania; and a magnificent display occurred on the 7th of January, 1831, which was visible at Lake Erie, and throughout northern and central Europe. Another aurora, that happened on the 3d of September, 1839, was seen at the Isle of Sky, 57° 22′ N. Lat., at Paris, New Haven, and at New Orleans.

570. The beautiful phenomenon of the northern light is not confined to the northern hemisphere. An *aurora australis*, or *southern light*, was observed by Don Ulloa, at Cape Horn, in 1745; and various displays were seen by Capt. Cook, in the high southern latitudes, at the same time that the *northern lights* were visible in Europe.

In the late Exploring Expedition, during the southern cruise of the Peacock and Flying Fish, several brilliant auroras were seen, which are thus recorded. On the 18th of March, 1839, there was "a beautiful display of the *aurora australis*, extending from S. S. W. to the east; the rays were of *many colors*, radiating towards the zenith and reaching an altitude of 30°. On the 19th, in about 68° S. Lat., another display was witnessed which exhibited a peculiar effect. In the southern quarter of the heavens there was the appearance of a *dense cloud*, resembling a shadow cast upon the sky, and forming *an arch* about 10° in altitude. Above this were seen coruscations of light, rendering all objects around the ship visible. From behind this cloud, diverging rays frequently shot up to an altitude of from 25° to 45°. These appearances continued until the day dawned."

What is said of the extent of the aurora?
Are auroras seen in the southern hemisphere?
What are they called?
Give instances.

571. HEIGHT. The height of the aurora has been variously estimated The earlier philosophers computed its altitude at several hundred miles; but a much lower limit is assigned by later observers. An aurora which appeared in March, 1826, at different places in England, was calculated by Dr. Dalton to be 100 miles high. Observations for determining the elevation of the splendid aurora of January 7th, 1831, were made by Christie and Hansteen, but their computed heights varied from 23 miles to 120.

In the brilliant display that happened on the 14th of November, 1837, the estimated altitudes were even more discrepant, varying from one to two hundred miles.

A very distinct auroral arch was seen at various places throughout the Northern and Middle States, at about ten o'clock on the night of the 7th of April, 1847. From the observations taken by Mr. E. C. Herrick, at New Haven, Ct., and Dr. P. W. Ellsworth, at Hartford, Ct., the height was computed by the former gentleman, and found to be *one hundred and ten miles.* These observations having been made under favorable circumstances, and being accordant with each other, this result is entitled to great confidence.

The height of the northern lights is obtained in the way that has been already described; but such is their fitful nature and varying form, that two distant observers can scarcely ever be sure that they have measured the angular height of the same part of the aurora. Hence arise these discordant calculations upon the same phenomenon.

572. There is every reason for believing, that the auroral light is at times *very near* the earth, and even within the *region of the clouds.*

During the polar expedition of Franklin, in 1820, observations were taken by Hood and Richardson, upon three auroras, at stations eighteen leagues distant from each other, and the heights which they obtained, were found to vary from *six* to *seven* miles; while an aurora

---

Relate in full the calculations respecting the height of the northern lights.

beheld by Farquharson, of Scotland, was computed to be as low as 4300 feet. Franklin thus remarks: "The fact that the aurora exists at a less height than that of dense clouds, was evinced at Fort Enterprise, on *two* or *three occasions*, during the night of the 13th of February, 1821, and particularly about midnight, when a *brilliant mass of light, variegated with the prismatic colors, passed between a uniformly steady, dense cloud and the earth.* In its progress, that portion of the cloud which the stream of light covered was *completely concealed* until the coruscation had passed over it, when it appeared as before."

573. A similar, but more extraordinary phenomenon, which occurred during his third Arctic voyage, is thus related by Capt. Parry. "While Lieutenants Sherer, Ross, and myself were admiring the extreme beauty of the northern lights, we all simultaneously uttered an exclamation of surprise, at seeing a *bright ray of the aurora* shoot suddenly downward from the general mass of light, and *between us and the land*, which was there distant only *three thousand yards*. I have *no doubt*, that the ray of light *actually passed within that distance* of us."

574. SOUNDS ATTENDING THE AURORA. It has been asserted, that the aurora is sometimes accompanied by *a noise* like the *rustling of silk*, or the *sound* of a *fire* when excited by the wind; but much difference of opinion has arisen upon this point. Those who are incredulous in this particular, affirm that the noise in question may be nothing more than the murmur of the ocean, or of the forest; the rustling of the snow as it is driven by the wind, or the crackling sound that arises from its freezing; all which, it is said, might be easily attributed to the aurora, when the mind is excited by the wondrous spectacle, and susceptible to every illusion —the splendors that burst upon the sight, and the sounds which strike the ear being then referred to the same origin.

State the facts showing that the aurora is at times very near the earth. Give the facts respecting the sounds attending the aurora.

575. Scoresby, Richardson, Franklin, Parry and Hood, during their Polar expeditions, *never heard any sound* which they considered as proceeding *undeniably* from the northern lights, though *hissing noises* were heard during the auroral displays which were attributed by them to one or more of the preceding causes. These observers do not, however, *deny*, that at times *audible sounds* proceed from the aurora, and even express such a belief, founded upon the concurrent testimony of the natives of the arctic climes.

576. Credible observers in Iceland, Siberia, and Scandinavia, have *never heard* these singular sounds; nor were they perceived by the French scientific expedition, which wintered at Bossekop, in 1838–39; but Hansteen claims to have established their existence from a series of observations in the high northern latitudes. Upon this subject, Simpson thus remarks in his Northern Discoveries when speaking of a brilliant aurora seen by his attendant, at Fort Confidence, on the 5th of March, 1839, "'The aurora seemed to ascend and descend, accompanied by an *audible sound* resembling the rustling of silk. This lasted about *ten minutes*, when the whole phenomenon suddenly rose upwards, and its splendor was gone. Ritch is an intelligent and credible person, and on questioning him closely, he assured me that he had perfectly distinguished the *sound* of the *aurora* from that produced by the *freezing of the breath*, for the temperature was forty-four degrees below zero. I can therefore no longer entertain *any doubt* of a fact *uniformly asserted by the natives*, and insisted on by my friend Mr. Dease, and by many of the oldest residents of the fur countries, though I have not had the good fortune to hear it myself."

577. TIME. The appearance of the northern lights is not confined to any particular hour of the night, a fact which is fully proved by the circumstance that the *same display* is frequently witnessed at places widely differing in longitude. Thus, if the aurora extends

from Boston, Mass., to Berlin, in Germany, and is beheld simultaneously at these cities, the difference in the reckoning of time will be nearly *five hours and a half* (C. 939).

578. There is much reason for believing that the aurora sometimes occurs during *the day*, though rendered invisible by the presence of the sun. Richardson perceived at Bear Lake, the motion of the aurora before the entire disappearance of twilight, and even during the day he discerned *clouds*, arranged in *columns and arches*, resembling those of the northern lights. Besides, as we shall show hereafter, a brilliant display of this phenomenon is always accompanied by a greater or less *disturbance* of the *magnetic-needle*, (C. 997,) and as these disturbances take place in the *day* as well as in the *night*, it is reasonable to infer that they are caused by the presence of an invisible aurora.

579. FREQUENCY. This phenomenon is more frequently seen in *winter* than in *summer;* we must not, however, hastily conclude from *this circumstance*, that the number of auroras during the former season is *actually greater*, for the *increased length of the nights during the winter* would enable us then to see more displays of the northern light, even if the *times* of its occurrence were equally distributed throughout the year. About the period of the equinoxes they also appear to be more frequent. These facts are shown from the following table of Kaemtz, which gives the number of auroras that have been seen in each month.

NUMBER OF AURORA BOREALES IN EACH MONTH.

January,	229.	July,	87.
February,	307.	August,	217.
March,	440.	September,	405.
April,	312.	October,	497.
May,	184.	November,	285.
June,	65.	December,	225.

Why is it supposed sometimes to occur in the day?
What is said respecting the frequency of its appearance in winter and summer? Recite the table.

580. In addition to this annual variation, there appears to be another which extends through a considerable number of years, but of which very little is known. Thus, from 1707 to 1752, the northern lights became more and more frequent; but after the latter date, a period of twenty years occurred, in which they diminished in number.

An increase in their frequency began in 1820, and since that period many magnificent displays have been witnessed.

The number observed for the last ten years, at New Haven, Ct., by Mr. E. C. Herrick, is shown in the following table.

				Number of Auroras
From May,	1838,	to May,	1839,	35.
"	1839,	"	1840,	36.
"	1840,	"	1841,	36.
"	1841,	"	1842,	21.
"	1842,	"	1843,	7.
"	1843,	"	1844,	7.
"	1844,	"	1845,	12.
"	1845,	"	1846,	19.
"	1846,	"	1847,	20.
"	1847,	"	1848,	28.

Between the 12th of September, 1838, and the 18th of April, 1839, no less than *one hundred and forty-three* distinct auroras were seen by the French observers at Bossekop. They were most frequent at the period when the sun was below the horizon, viz.: from the 17th of November to the 25th of January. During this night of *ten weeks, sixty-four* auroras were visible.

581. DISTURBANCE OF THE MAGNETIC-NEEDLE. During the prevalence of the aurora, the *compass-needle*, instead of remaining *motionless*, in the magnetic meridian, is *often much disturbed.* Sometimes it is *deflected* toward the *east* several minutes and even degrees; then

---

Is there any other probable variation?
Recite the table.
What is said respecting the disturbance of the compass-needle?

it is agitated, and returns either slowly or rapidly, to the *meridian*, which it passes at times and moves toward the *west*. These deviations are as changeable as the phenomenon itself. When the *arch* is *motionless* the *needle* is *quiet*; its disturbance *commences* when the streamers begin *to play*.

582. Franklin observed at Fort Enterprise, that the disturbance of the needle was *simultaneous* with some *change* in the *form* or *action* of the northern lights, and that after being deflected it returned to its former position *very gradually*, not resuming it before the *following morning*, and sometimes even not *before noon*. Moreover when the *auroral arch* was either at *right angles* to the meridian, or its *western extremity north* of *west*, the needle was *deflected toward the west;* but if its *western extremity* was *south of west*, the needle *moved toward the east*.

During the aurora of November 14th, 1837, the entire range of the needle at New Haven, was observed by Messrs. Herrick and Haile to be nearly *six degrees*. It was not until the morning of the next day, between seven and nine o'clock, that the needle was at rest in its usual position.

583. This effect upon the magnetic needle during the prevalence of the northern lights, was noticed for the first time by Celsius and Hiorter, at Upsal, on the 1st of March, 1741.

584. It is asserted by Wilke, that when the aurora appears in great splendor, the position of the *dipping-needle* is as variable as that of the *compass-needle;* the former *rising* and *falling* with the northern crown.

Hansteen has also observed, that the dipping-needle descends very much *below* its usual position before the aurora is visible; but that after the display commences it begins to *rise:* and more rapidly in proportion to its brightness. The needle then *slowly* resumes its original position, which it frequently does not attain until

---

How great was its range at New Haven, November 14th, 1837?
What has been observed respecting the dipping-needle?

twenty-four hours have elapsed. From numerous observations at Bossekop, M. Bravais has likewise obtained the same results.

585. CAUSE. No satisfactory explanation has ever been given of this singular phenomenon: that a connection exists between the aurora and the magnetism of the earth, is evident from the preceding facts; but the *nature* of that connection is still unknown.

To trace all the hypotheses which have been started would be an unprofitable task; but a glance at some of the most prominent may be given. Canton supposes the aurora to be caused by the passage of electricity from *positive* to *negative* clouds, in the upper and rarefied regions of the atmosphere. He adduces in support of this view the fact, that when the air within a long, glass tube is rarefied, and electricity passed through it, the whole tube is illumined by *flashes of light* traversing its entire length. It may, however, be stated in reply, that the general height of the northern lights far exceeds that of the highest clouds.

586. Beccaria supposes, that there is a constant circulation of the electric fluid from north to south, and that the aurora is seen, whenever the electrical current passes *nearer than usual* to the earth, or the state of the atmosphere is such as to render it *luminous*. Faraday has demonstrated, that the electricity of the earth necessarily tends from the *equator towards the poles;* and has suggested, that the aurora may possibly arise from an *upward current* in the atmosphere flowing back from the poles towards the equator.

Kaemtz conjectures, that since a spark is perceived every time an electric current produced by a magnet is *broken*, the northern lights may perhaps be caused by a rupture in the magnetic equilibrium of the globe. At the same time, however, he utterly disclaims the idea of accounting for all the circumstances of this wonderful phenomenon, in our present imperfect state of knowledge.

What is known of the origin of the northern lights?
State the hypotheses given.

587. UTILITY. The light of the aurora, from its frequency and splendor, serves materially to relieve the darkness and enliven the gloom of the long polar night. During this period, its play is almost incessant, (Art. 580,) and its coruscations exceedingly vivid and beautiful.

So brilliant is the aurora in these regions, that Maupertius and others, who were sent to Lapland in 1735, by the Academy of Sciences of Paris, for the purpose of measuring an arc of the meridian, were enabled to pursue their difficult work by the light it afforded, long after the sun had ceased to be visible. And Maupertius remarks, that its light, together with that of the moon and stars, is sufficient, during this season, for most of the occasions of life.

What useful purpose does the aurora subserve in the polar regions?

PART VII.
MISCELLANEOUS PHENOMENA.

CHAPTER I.
OF THE FALL OF TERRESTRIAL SUBSTANCES FOREIGN TO THE ATMOSPHERE.

588. In addition to storms of rain, hail, and snow, which are products *peculiar* to the atmosphere, and are the results of the operations of well-known agencies and laws, showers of matter of a *terrestrial* nature have not unfrequently occurred, which have been traced, upon close examination, to a *mineral, vegetable,* and even *animal* origin. The most remarkable of these singular phenomena are *dust-storms* and *blood-rains*, which will now be described.

DUST-STORMS AND BLOOD-RAINS.

589. From time to time, and in regions of the globe widely separated from each other, *dust* in large quantities has descended from the heights of the atmosphere, not only upon the *land*, but also far out on the *ocean, hundreds of miles* from the shore. It is entirely distinct from that which is disseminated through the air by the winds, during the eruption of volcanoes, and for many years has been described, by observers and writers, under the various names of *dust-storms, dust-rain, red fogs, Sirocco dust,*

What is the subject of part seventh?
Of what does this chapter treat?
In addition to storms of rain, hail, and snow, what other kinds of showers have not unfrequently happened?
What are the most remarkable of these phenomena?
What is said respecting the *fall of dust* from the heights of the atmosphere?
What are the various names under which this phenomenon has been described?

African dust, sea-dust, Atlantic dust, and *tradewind-dust.*

590. This dust not only falls dry, in the form of a fine, *impalpable powder,* but is occasionally mingled with *rain, hail,* and *snow,* which it dyes with its own hue. As it is usually of a *reddish* color, these showers of rain and storms of hail and snow have received the appellation of *blood-rains.*

DUST-STORMS.

591. INSTANCES. On the 20th of October, 1755, a *black dust,* like lamp-black, fell in Shetland, between 3 and 4 o'clock in the afternoon. The sky at the time was hazy, and the dust fell in such quantities as to cover the hands and faces of persons exposed to it, and to blacken their linen.

592. During the 5th and 6th of March, 1803, while the wind was blowing from the *south-east,* a shower of *red dust* fell in Italy. Ten years afterwards, on the 14th of March, 1813, a similar storm occurred at the town of Gerace, in Calabria. According to Prof. Sementini, of Naples, the wind, in the early part of the day, blew from a western quarter, bringing up dark, heavy clouds from the sea over the land.

At about 2 o'clock in the afternoon the wind subsided, while a deep gloom pervaded the air, and the clouds grew *red* and threatening. Thunder followed, and soon after *red dust,* mingled with *red rain and snow,* descended upon the town. This dust had the appearance of a *fine powder.*

593. A shower of dust fell at Malta on the 15th of May, 1830, and at the same time a similar fall occurred in the bay of Palmas, in Sardinia, while a Sirocco wind was blowing from a south-easterly quarter. The Maltese dust was of a *brownish-red hue.* Some of it was collected by Mr. R. G. Didman, of the ship Revenge, and for-

---

What are *blood-rains*, and why are they so called?

Relate the various instances given of dust-storms in Shetland, Italy, Gerace, Malta, and Genoa.

warded to Mr. Charles Darwin, an eminent English naturalist, for examination.

594. On the 16th of May, 1846, a shower of Sirocco-dust occurred at Genoa, having the same *brownish-red* hue as the dust which fell at Malta in 1830.

Six months afterwards a remarkable storm of this nature swept over Lyons, in France, and so thickly did the dust descend, that the amount which fell at this time was computed to weigh no less than *thirty-six tons*.

595. In the year 1831, the ship Beagle, under the command of Captain Fitzroy, was dispatched by the British government on a voyage of scientific discovery around the world. Mr. Darwin, the naturalist just mentioned, accompanied the expedition, and during the voyage observed a *dust-shower*, near St. Jago, the chief of the Cape de Verd isles.

The morning before the Beagle anchored at Port Praya, in St. Jago, Mr. Darwin collected a little package of impalpable *brown*-colored dust, which appeared to have been filtered from the wind by the gauze of the vane at the mast-head. In speaking of this phenomenon, he remarks, that the atmosphere in this region is usually filled with a *haze*, caused by the falling of this fine, *brown-colored dust*. By the kindness of a friend, Mr. Darwin received *four parcels of dust* which fell upon the deck of a vessel, a few hundred miles north of the Cape de Verd isles.

596. Much valuable information respecting dust-showers on the ocean has been gathered by this gentleman, who has found fifteen different accounts of the descent of dust upon ships when far out on the Atlantic. It has often fallen upon them when they were *several hundred*, and even a *thousand miles* from the coast of Africa, and at points *sixteen hundred miles distant* in a *north* and *south direction*.

597. In some of the dust collected upon a vessel *three hundred miles* from land, particles of stone were discov-

---

State what is said respecting the fall of dust on the ship Beagle.

What is known, from the researches of Mr. Darwin, in regard to the Atlantic dust?

ered, more than the *thousandth of an inch* square, mixed with finer matter. It falls in such quantities as to soil every thing upon which it descends, and to irritate the eyes of persons exposed to it. Ships have even been known to run ashore, owing to the obscurity of the atmosphere resulting from the presence of this dust.

598. The occurrence of *dust-showers* in the vicinity of the Cape de Verd isles has been noticed, at intervals, from the year 1579 to the present time. The extent of the region over which they here prevail varies, according to Darwin, from 960,000 to 1,280,000 square miles; but a greater estimate is given by Captain Tuckey, who supposes that it ranges from 1,648,000 to 1,854,000 square miles. The Atlantic dust is believed by Mr. Darwin to come from Africa, since not only does wind blow from that quarter whenever it falls, but the showers also occur during those months when the harmattan is known to raise clouds of dust high into the atmosphere.

599. During a voyage from Richmond, Va., to Rio Janeiro, in the winter of 1845–6, Mr. Thomas Ewbank, of the U. S. Patent Office, met with many instances of the falling of *sea-dust*, and traced the rich and peculiar hues, that at times adorned the clouds and sky, to the diffusion of this fine powder throughout the intermediate atmosphere.

600. On the 10th of January, 1846, in 23° 33′ N. Lat., and 34° 37′ W. Long., he observed a narrow belt of *slate-colored* sky skirting the horizon, while upon this rested a broad band of *vermilion*, interspersed with soft dashes of *Indian ink*, shaded with *umber*. These hues changed, by insensible degrees, into a bright *cream-color*, and this again into a pale, delicate *green*, which deepened in tint as it approached the zenith, while over all floated *amber-colored* clouds, growing richer in hue and smaller in size as they sunk towards the horizon.

What is the extent of the region over which the Cape de Verd and Atlantic dust-storms prevail?

What is the opinion of Mr. Darwin as to the origin of this dust?

Relate in full the account given by Mr. Ewbank of the dust-storms that he observed on a voyage from Richmond to Rio Janeiro.

601. Three days afterwards, in 16° 07' N. Lat., and 31° 13' W. Long., the wind blew strongly *from the east*, bearing along with it a *red, impalpable powder*. This minute dust was seen on the windward side of the sails, where it was supposed to have been collecting during the two previous days. It was extremely fine, and could only be seen by bringing the loose fibres of a rope, upon which it had settled, between the eye and the sun, when its presence and color were readily discerned.

602. The sun throughout the day, as well as the moon at night, was enveloped in a *haze*, which was supposed to be caused, in some measure, by the dust that floated in the air. The captain of the vessel, who had noticed this phenomena before, called the red powder *African sand*.

603. During the two following days the heavens presented scenes of gorgeous and surpassing beauty, the colors of the sky and clouds ranging through *emerald green*, *pink*, *purple*, *crimson*, *yellow*, *chocolate*, *umber*, and *slate;* while beneath this rich and varied combination a groundwork of the purest *cream-color* extended, giving tone to the whole, and changing in tint from a *fawn-color* to a *pale white*.

604. On the 16th of January, in 7° 44' N. Lat., and 28° 31' W. Long., the *red dust* was observed to *accumulate* upon the vessel—an old sail, looking as if it had been painted of a *light brick color*. The ship at this time was opposite Soudan and Senegambia, which border on the great African desert, whence the captain supposed the shower to come. A portion of the dust was collected by rubbing a piece of foolscap paper over the colored sail.

605. A *fall of dust*, accompanied by snow, occurred in the month of February, 1850, at Olsterholz, near Detmold, in Westphalia. The wind, during this phenomena, blew from the south-west. The dust fell so thickly as to cover the earth to the depth of *one eighteenth of an inch*.

BLOOD-RAINS.

606. INSTANCES. On the 12th of August, in the year

What is said of the dust-shower that occurred at Olsterholz?

1222, a *red rain* fell at Rome for the space of a *day and a night;* and a similar event occurred at Cremona on the 3d of July, 1529. In 1608 a *red rain* descended for several miles around Aix, in France; and in 1623 another blood-rain happened at Strasburg, between 4 and 5 o'clock in the afternoon. On the 5th and 6th of May, 1711, *red rain* fell at Orsio, in Sweden; and a shower of this nature also occurred near Genoa in the year 1744.

607. A very remarkable rain of this character fell at Locarno, in Switzerland, on the 14th of October, 1755. A warm Sirocco wind was here blowing at 8 *o'clock* on the *morning* of this day, and *two hours* afterwards the air was filled with a *red mist*.

At 4 *o'clock* in the afternoon a *blood-rain* descended, which left on the ground a *reddish deposit*. *Nine inches* of this colored rain fell, in the course of one night, over a region *forty square German leagues* in extent. It even reached Suabia, on the northern side of the Alps; while amid the cold heights of these lofty mountains it changed into a reddish snow, which fell to the depth of *nine feet*.

608. The *red matter* that was deposited during this shower was found, by actual measurement, to be in some places an *inch deep*, or *one-ninth* part of the quantity of rain. Upon the supposition that it fell, on an average, to the depth of only *one-sixth* of an inch, *twenty-seven hundred cubic* feet of this red substance must have covered *every English square mile*.

609. On the 13th of November, 1755, a *red rain* fell in Russia, Sweden, Ulm, and on the Lake of Constance; and on the 9th of October, 1763, a similar shower descended at Cleves, Utrecht, and many other places in Europe.

610. During the remarkable phenomenon that occurred at Gerace, on the 14th of March, 1813, the *red rain* prevailed over a great extent of country, falling throughout the two Calabrias, and on the opposite side of the province of Abruzzo, in the kingdom of Naples.

Relate, in detail, the several instances given of the fall of blood-rain.

611. A *red rain* likewise fell at Sienna, and upon the adjacent country, on the 15th of May, 1830, at 7 o'clock in the evening, and also at midnight. The weather for two days previously had been calm, but the sky was overcast with dense, *reddish* clouds.

612. BLACK RAIN. The material that mingles with these extraordinary rains is not always of a red hue, but is sometimes of a *dark color*, and imparts an *inky blackness* to the shower. A rain of this kind occurred at Montreal, in Lower Canada, on two several days during the month of November, 1819, under the following circumstances:

On the morning of the 21st of this month a dense gloom enveloped the city, while the whole atmosphere was obscured by a *thick haze*, of a *dusky orange color*, and at this time rain descended of a *dark inky hue*. The weather soon after became pleasant, and continued so until the following Tuesday, when at noon the whole city was again shrouded in a *heavy, damp vapor*, so dense that it became necessary to light candles in all the houses.

At about 3 o'clock in the afternoon a slight shock of an earthquake was felt, attended by a noise like the discharge of distant artillery.

Soon after, when the darkness was the deepest, the gloom was dispelled by a vivid flash of lightning, which was followed at once by a crashing peal of thunder; and this was succeeded by a heavy shower of *thick, black rain*.

613. On the 22d of April, 1846, a copious *black rain* fell also in England, in the towns of Dudley, Stourport, Abberly, and Bewdley, which are situated in the northern part of Worcestershire.

This shower lasted from 11 o'clock in the morning till 1 o'clock in the afternoon, the rain descending so abundantly as to *blacken* the waters of the places where it fell, and darken the river Severn.

Give an account of the *black* rain of Montreal.
Of that which happened in Worcestershire.

614. RED HAIL. A storm of *red hail* is stated by Baron Humboldt to have once occurred at Paramo, in South America, between Bogota and Popayan. There likewise fell over all Tuscany, on the 14th of March, 1813, a shower of hail of an *orange hue*.

615. BLACK HAIL. A hail-storm happened in Ireland on the 14th of April, 1849, which deposited upon the ground a *black, inky* substance. Some of this dark matter was collected and examined, and found to be of the same nature as the coloring material of *red rains*.

STORMS OF COLORED SNOW.

616. RED SNOW. One of the most remarkable falls of *red snow* on record is that which has already been mentioned (Art. 607), as occurring simultaneously with the blood-rain of Locarno, in Switzerland, when snow of a *reddish* hue covered the neighboring Alps to the depth of *nine feet*.

617. On the 5th and 6th of March, 1808, *red snow* fell for the space of three nights in Carniola, a province of Germany, and throughout Carnia, Cadore, Belluno, and Feltri, to the depth of *five feet* and *ten inches*.

The earth had been previously covered with *white snow*, and the storm of colored snow was succeeded by another, the flakes of which were as usual, of a pure and brilliant white. The two kinds were perfectly distinct. When a portion of the red snow was melted in a vessel, and the water evaporated, a fine *rose-colored, earthy sediment* remained at the bottom. Red snow, likewise, fell at this time on the mountains of the Valtelline, in Switzerland, at Brescia, and on the Tyrol.

618. During the *dust-shower* and *blood-rain*, at Gerace, *red snow* descended over a wide extent of country, embracing the two Calabrias, Tolmezzo, and the Carnian Alps. In Tuscany it fell of an *orange hue*, while at Bologna its tint was a *brownish yellow*.

What instance is given of the occurrence of *red* hail?
What of *black* hail?
Where, when, and under what circumstances have storms of colored snow occurred?

619. On the 15th of April, 1816, *colored snow* fell in Italy, upon Tonal, and on other mountains. It was of a *brick-red hue*, and, when melted and evaporated, a light and impalpable *earthy powder* remained.

620. A storm of *reddish snow* took place on the 31st of March, 1847, in Puster Valley, in the Tyrol. It derived its tint, which was a *brownish red*, from a fine *colored* dust, resembling that of the Atlantic showers.

621. BLACK SNOW. A few years ago a fall of *black snow* occurred in New Hampshire, at Walpole, and the adjoining towns. A person writing to the Boston Journal from Walpole, remarks, in relating this extraordinary phenomenon : " I send you some writing, written with the *snow as it fell*, and with a *clean pen*." This writing, according to the editor of the Journal, was *perfectly legible*, and appeared as if having been written with *pale black ink*.

622. These colored snows must not be confounded with those already described in Arts. 286; 287, and 288. The snows there mentioned are *white before their fall*, and acquire their red and green tints, *after their descent*, from the presence of a microscopic plant whose cells are filled with animalcules, and which, even in Arctic climes, spreads itself with extraordinary vigor over fields of snow. On the contrary, in *storms of colored snow*, the coloring matter is *in the atmosphere*, and the snow is *dyed before its fall*.

NATURE OF THE DUST.

623. It appears from the microscopic investigations of eminent observers, and especially from those of Ehrenberg, that the dust which causes dust-storms, and produces the phenomenon of blood-rains, is composed both of *organized* and *unorganized matter :* the *latter* being *portions of various minerals*, while the *former* consists principally of the *shells* of *infusoria*, mingled with fragments of *petrified plants* and *parts of insects*.

Are these colored snows the same in character as those already described in Arts. 286, 287, and 288 ? Why not ?
Of what is the dust of *dust-storms* and *blood-rains* composed ?
Of what does the organic matter consist ?

It may not be amiss to explain to the student in this place the meaning of the term *infusoria*.

624. INFUSORIA. The general name of *Infusoria* has been given to those minute living beings which can only be seen by the aid of the microscope. On account of their being first detected in vegetable *infusions*, they are termed *infusoria;* and since they are exceedingly small, they have also received the appellation of *animalcules*, or *little animals*. They are found in countless myriads in all waters, and in the fluids that circulate in animal and vegetable bodies, while their shells and eggs are disseminated by the winds over every part of the world.

625. More than *eight hundred* distinct species have been discovered, possessing the most grotesque and singular forms. Some resemble globes, trumpets, stars, boats, and coins; others assume the forms of eels and serpents, and many appear in the shape of fruits, necklaces, pitchers, wheels, flasks, cups, funnels, and fans.

Their minuteness is almost incredible, for the monad, the smallest of all living beings, never exceeds in length the *twelve thousandth part of an inch*. A single shot, *one-tenth of an inch in diameter*, occupies more space than *seventeen hundred millions* of these atoms—each in itself a perfect being, amply endowed with vital powers adapted to the mode and range of its existence.

626. STRUCTURE. The *outer covering* of the infusoria is of *two* kinds; the *first* is soft and yielding, resembling the skin of the leech and slug; but the second is a fine, *transparent shell*, possessing a flexibility like horn. Those animalcules that are protected by the latter integument are termed *loricated*, from the Latin word *lorica*, a *shell;* while the name *illoricated*, or *shelless*, is assigned to those which are invested with the softer covering.

The material that composes the shells varies in differen: species. In many kinds it consists entirely of *flint*, and

---

Describe the infusoria—the number of their species—their minuteness.

What is said respecting their structure?

in others of *lime,* united with *oxide of iron.* In some cases it is combustible.

627. When the loricated infusoria die, their shells remain undecayed for ages, often congregated in such countless myriads as to form large portions of the earth's surface.

The city of Richmond, in Virginia, is built upon an extensive bed of *flinty marl,* from *twelve* to *twenty feet* in thickness, filled with *fossil, infusorial shells;* and it is stated by geologists, that nearly half of the bulk of all the chalk of Northern Europe is composed of the *fossil remains* of animalcules, and other minute shells. They are mingled with the mud that forms the bed of the Arctic Ocean; they float with the iceberg in all its wanderings, and lie loosely scattered over the surface of every land. These hieroglyphics of nature are interpreted by the aid of the microscope.*

628. THE ITALIAN DUST-SHOWER OF 1803, AND THE CALABRIAN OF 1813. In the dust which fell in Italy during the month of March, 1803, forty-nine species of organic structures were discovered, and sixty-four in that which descended at Gerace, in Calabria, in 1813. Thirty-nine species in the Italian dust-shower, and fifty-one in the Calabrian, are *identical* with those discovered in more recent dust-storms. It is worthy of remark, that these two storms, though *ten years* apart, have no less than twenty-eight species in common, and in *both nearly all* the species are of *fresh-water origin.* Among the numerous infusorial shells, *four South American forms* were discovered; of these, *one* occurs in Peru, *another* in Surinam, and the remaining *two* belong to Chili. *No* animalcular structures were found exclusively African.

629. ATLANTIC AND CAPE DE VERD DUST. The dust

When the loricated infusoria die, what becomes of their shells?
Of what did the dust consist which fell in the Italian and Calabrian dust-storms?

---

* For further information on the subject of Living and Fossil Infusoria, see "Views of the Microscopic World," by the author; published by Farmer, Brace & Co., New York.

that was collected by Mr. Darwin on the Atlantic, in N. Lat. 17° 43′, W. Long. 26°, and at the distance of about *five hundred miles* from the African coast, was submitted to the examination of Ehrenberg, who discovered that *one-sixth* part of it was composed of the *flinty shells* of *fresh water* and *land* infusoria, and of *silicious fossil plants*. There were eighteen species of the former, and as many of the latter. Of the animalcular remains, the greater part were European; *one species* was *decidedly* of South American origin, and *another* probably; but there were *none* that belonged exclusively to Africa. In the opinion of Ehrenberg, the two South American species were either brought from that country by the upper winds of the atmosphere, or from some other locality which is yet unknown.

630. In the dust of several other showers, which occurred between the years 1834 and 1838, some at St. Jago, and some on the neighboring ocean, numerous organized structures were discovered, thirty of which were *different* from those detected in the dust just described. Among these were the shells of a few South American infusoria, and one beautiful microscopic shell, termed the *Polythalamia*,* or *many-chambered shell*. A single species was observed that occurs in the Isle of France; but *none* of the forms were recognized as *peculiarly African*.

631. Some of the dust collected by Mr. Ewbank, on his voyage to Rio Janeiro, was examined by Professor Bailey, of West Point; but he was unable to discover in it any thing besides irregular, *inorganic*, mineral fragments. He believes, however, that more interesting results would have been obtained if the dust had been gathered with greater care. The entire number of *distinct* organic

---

Relate, in full, what is said respecting the composition of the Atlantic and Cape de Verd dust.
What is said respecting the dust of several other showers?
Were any of the forms distinctively African?
Were any organisms discovered in the dust collected by Mr. Ewbank?
What is, however, the opinion of Prof. Bailey?

---

* From *polus*, (Greek,) many, and *thalamus*, (Latin,) a chamber.

forms hitherto discovered in the Cape de Verd and Atlantic *dust-storms* is *sixty-seven*.

632. SIROCCO DUST. The dust that fell at Malta on the 15th of May, 1830, afforded forty-three distinct organized forms; of these there were fifteen *infusorial structures*, twenty-one kinds of *minute, petrified plants*, and seven of *Polythalamia*. The species of animalcules were, for the most part, identical with those discovered in the Cape de Verd and Atlantic dust-showers.

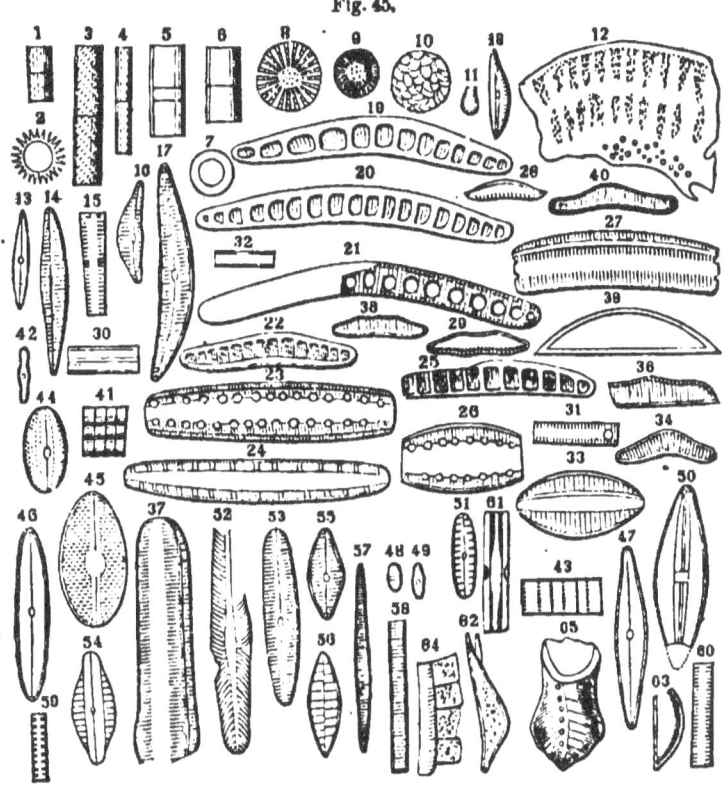

Fig. 45.

MICROSCOPIC ORGANISMS OF THE LYONS DUST-SHOWER.
(*Fossil Infusoria.*)

One form was noticed belonging peculiarly to Chili, but *none* were found distinctively *African*.

State the entire number of organic forms hitherto detected in the Cape de Verd and Atlantic dust-storms

254   MISCELLANEOUS PHENOMENA.

633. The Sirocco dust that fell at Lyons, on the 17th of October, 1846, was so rich in organic remains that they constituted *one-eighth* part of its mass. They consisted of numerous species of *infusoria* and of *petrified plants*, mingled with a few kinds of *Polythalamia*, and minute, vegetable fragments. The species were nearly all of *fresh-water* origin, *one-seventh* only being *marine*. In figures 45 and 46 are delineated the various microscopic

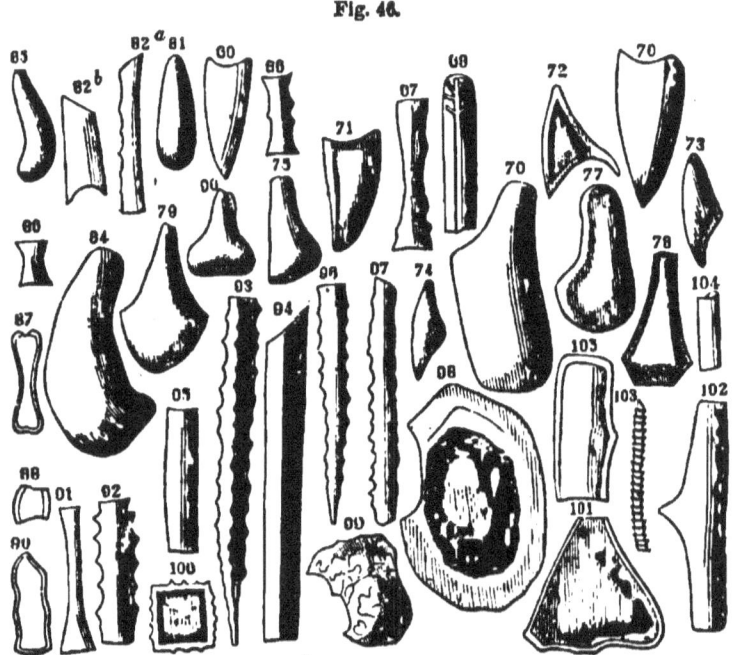

MICROSCOPIC ORGANISMS OF THE LYONS DUST-SHOWER.
(*Fossil Plants.*)

organisms which were discovered in this dust. The most remarkable circumstance respecting it is the fact, that, notwithstanding its general resemblance to the dust of the Atlantic showers, which has always exhibited nothing but *dead* and *empty* infusorial shells, this, on the contrary, was found, in many cases, to contain a species of

---

Describe the nature of the Sirocco dust that fell at Malta and at Lyons.
What is remarkable respecting the Lyons shower?

infusoria which was distinctly seen to be filled with *green ovaries*, or egg-sacks, and consequently *was capable of life*.

634. The dust collected, in the preceding instances, from the Cape de Verd, Atlantic, and Sirocco showers, being *nine* in all, afforded 119 *distinct* organisms. Of these there were fifty-seven species of *infusoria*, and eight of *Polythalamia;* forty-six kinds of *fossil plants*, together with *particles* of seven kinds of *plants*, and one *fragment of an insect*. Only seventeen of these organisms were *marine;* while 102, *six-sevenths* of the whole, consisted of *fresh-water* species. In all these showers the dust exhibited *no indications* whatever of *volcanic* origin.

635. In three dust-showers which occurred in the years 1847 and 1848—the *first* in Salzburg, the *second* in Arabia, and the *third* in Silesia and Lower Austria—similar fresh-water organisms were detected. The same South American species were here found, as in other showers, without any characteristic African forms.

636. The red snow that fell in the Tyrol on the 31st of March, 1847, afforded sixty-six different organic forms. Of these, twenty-two were *infusorial structures*, twenty-eight *fossil plants*, two *polythalamia*, and thirteen *particles of plants*. There was also one *fragment* of *an insect*. The greater part, by far, of all these species, were of *land origin*, *two* only being marine.

A remarkable resemblance exists between the coloring matter of this shower and the dust of the Atlantic, Genoese, and Lyons storms, not only in its *hue*, but in its *composition;* for out of these sixty-six structures, *forty-*

How many distinct organisms were discovered in the dust of nine showers? Describe the several kinds.

Was there any trace of volcanic dust in these showers?

What is observed respecting the dust-storms which happened in the years 1847 and 1848?

What organisms were detected in the red snow of the Tyrolese storm?

What resemblance was observed between the coloring matter of this shower and the dust that fell on the Atlantic, at Genoa, and at Lyons?

six are found in the Atlantic and Sirocco dust; and *twelve* species of infusoria and *twenty* of fossil plants are common to all.

637. In the dust that fell, mingled with snow, at Olsterholz, in the year 1850, Ehrenberg detected fifty organic forms, forty of which he had previously observed in the dust of other showers. The remaining ten species had never been before discovered in atmospheric dust.

638. NUMBER OF DISTINCT ORGANISMS DISCOVERED. In the dust of the various showers examined by this distinguished naturalist, no less than 320 distinct species of organisms were discovered. Of these, five only were of *marine* origin, and fourteen were forms peculiar to America.

639. NUMBER AND EXTENT OF DUST-STORMS AND BLOOD-RAINS. According to the researches of Ehrenberg, 340 instances of dust-storms and blood-rains are mentioned in history and in the annals of science, of which 81 took place before the Christian era, and 259 after it. These remarkable phenomena extend throughout the world, occurring on the ocean, on all the continents, and even in Australia. They appear, however, to prevail most within a zone, extending from that part of the Atlantic off the west coast of Middle and North Africa, along in the direction of the Mediterranean Sea, reaching a short distance north of this sea, and continuing into Asia between the Caspian Sea and the Persian Gulf, perhaps to Turkistan, Kaschgar, and even China: they seldom happen as far north as Sweden and Russia.

This zone, according to the observations of Captain Tuckey, has a breadth of 1800 *miles*.

What organic forms were discovered in the dust that fell at Olsterholz?

How many distinct organisms have been detected by Ehrenberg in the dust of numerous dust-storms? What is said respecting their origin?

What is the *number* and *extent* of *dust-storms* and *blood-rains*, according to Ehrenberg?

Where do they appear to prevail most?

What is the *breadth* of this zone?

640. THEIR PERIODICITY. These phenomena occur most frequently during the first half of the year; for out of 199 showers, whose dates are ascertained, 118 happened between January and July, and 81 between July and December. The distribution of the showers through the several months is as follows:

January,	27.	July,	9.
February,	14.	August,	17.
March,	23.	September,	7.
April,	18.	October,	18.
May,	18.	November,	16.
June,	18.	December,	14.

641. ORIGIN OF THE DUST. The *color* and *nature* of this dust; the circumstance that a *great quantity* of *earthy matter* sometimes falls in a *single shower*, as in that of Lyons; and the fact that dust-storms and blood-rains have occasionally happened from the time of Homer (900 B.C.) to the present day, have led Ehrenberg to advance a most extraordinary hypothesis. He believes that these phenomena are not to be traced to mineral matter belonging to the earth's surface; neither to masses of dust revolving in space, like the meteoric matter of Chaldni (Art. 555); nor yet to the influence of atmospheric currents, such as the trade-winds and harmattan, carrying the dust of the earth aloft into the air; but to some general law, as yet unknown, according to which *infusoria*, and other *living organisms, exist* and are *propagated* in the upper regions of the atmosphere.

The locality which constitutes the dwelling-place of these organisms he imagines to be of vast extent, and to be situated at the height of about 14,000 feet above the sea-level.

In what parts of the year do these phenomena most frequently occur?
How are they distributed through the months?
What are the views of Ehrenberg respecting the *origin* of the dust that falls in these singular storms?
Where does he suppose the abode of these organisms to be situated?

642. The apparent periodicity of the showers he accounts for by supposing that this cloud of organisms lies in the region of the trade-winds, and suffers partial and periodical deviations.

643. In the present imperfect state of our knowledge in regard to these phenomena, it would be highly unsafe to adopt this singular hypothesis.

Both the organic and inorganic matter contained in these storms are *terrestrial* in their nature, and the atmospheric currents are most probably the agents, which elevate this dust from the surface of the globe, and bear it along to distant regions.

644. The opinion, that the Atlantic, Cape de Verd, and Sirocco dust comes from the deserts of Africa, is inconsistent with certain known facts respecting it, and has therefore not been universally adopted. For instance, the *color* of this dust is *red*, while the *sand* of the African Saharas is *white* and *gray;* and we have also seen that none of the organized forms which it contains are peculiar to Africa; while many of them are *distinctively* South American.

645. It is the belief of Lieutenant Maury, that the red powder, which falls in these dust-storms, is brought by an upper wind from South America to Africa; where it descends and becomes the *lower* trade-wind, which disseminates the dust throughout the regions where it blows. It is not improbable that a portion of this dust, carried onwards by the higher current, falls within the sweep of the Sirocco—a circumstance which will fully explain the similarity that exists between the Sirocco and Atlantic dust.

---

* How does he account for the apparent periodicity of these showers and storms?

Are there, at present, sufficient reasons for adopting this hypothesis?

What is the *nature*, both of the *organic* and *inorganic* bodies, which constitute this dust?

*How are they probably raised into the atmosphere?

What reasons exist for believing that the Atlantic, Cape de Verd, and Sirocco dust does *not* come from Africa?

What is the opinion of Lieutenant Maury upon this point?

How can the similarity in the nature of the Sirocco and Atlantic dust be explained?

VOLCANIC SHOWERS.

646. The fall of ashes and dust soon after the eruption of volcanoes, is a phenomenon *entirely different* from *dust-storms* and *blood-rains;* for the materials which are precipitated in volcanic showers contain *no organic forms,* and are easily traced to their source.

647. CAUSE. The mighty energies that are at work, when a volcano is in full action, carry up the lighter portion of the ejected matter high into the air; it is then borne along by the upper winds, and at length falls, in showers, in regions often far remote from the burning crater.

648. INSTANCES—JORULLO. During the eruption of Jorullo, in Mexico, which began on the 28th of September, 1759, the sky was darkened with clouds of dust that afterwards fell at Queretaro, 100 miles distant; and during another eruption of the same volcano in 1819, dust, to the depth of *six inches,* descended in the streets of Guanaxuato, at the distance of 160 miles.

649. SOUFFRIERE. One of the most remarkable volcanic dust-showers on record is that connected with the eruption of the Souffriere mountain, in the island of St. Vincent, which occurred on the 30th of April, 1812.

650. On the 27th of this month the volcano, which had been slumbering for a hundred years, again burst forth, showering down sand, mixed with ashes and gritty, calcined particles of earth. This dust, driven before the wind, darkened the air like a cataract of rain, and covered the ridges, woods, and cane-lands with light grey-colored ashes, resembling snow. As the activity of the volcano increased, this continual shower extended farther and farther, destroying every trace of vegetation.

651. For three days the appearance of the burning mountain grew more awful and portentous, when at length, on the night of the 30th, a most terrific eruption

---

State what is said respecting *volcanic* showers—their cause.
Give an account of the showers attending the eruptions of Jorullo.
What remarkable shower of this kind is next mentioned? Describe it.

took place. From the midst of a lofty pyramid of flame issued streams of glowing lava, which, pouring down the sides of the mountain, flowed in torrents to the sea; while the sullen roar of these burning rivers was swelled by the thunderings and loud explosions of the crater. Stones, fire, ashes, and calcined masses rained down for hours, and earthquake following earthquake, almost incessantly, the whole island undulated like water shaken in a bowl.

652. On the next day, the air was so filled with volcanic dust, that it was dark at 8 o'clock in the morning; a dense haze shrouding sea and land. Most of the plantations in the vicinity of the Souffriere mountain were covered *ten* or *twelve inches deep* with dust and stones.

653. But the effects of this eruption were not confined to this island. During the night of the 30th the terrific explosions of the volcano were heard as far as Barbadoes, which is situated *seventy miles* due east from St. Vincent. On the next morning, at 4 o'clock, the atmosphere at Barbadoes was *bright and clear*: but at 6 o'clock the sky was obscured by thick clouds, from which issued in torrents, like rain, particles of volcanic matter finer than sand. At 8 o'clock, an appalling darkness, as intense as that which prevails in the depth of a stormy night, overspread the island, and continued till noon, but the showers of dust still descended at intervals until 7 o'clock in the evening.

654. This dust descended to the depth of *two inches*, and, according to the computation of observers, an average weight of 40,000 lbs. rested upon every acre on which it fell. Vessels at sea, some 300 miles, and others 500 to the windward of St. Vincent, had their decks covered with this volcanic dust. (Art. 103.)

655. TOMBORO. Still more surprising was the dust-shower, caused by an eruption of Tomboro, a volcano sit-

---

To what other island did this dust extend?
How far is it from St. Vincent?
How was the atmosphere of Barbadoes affected by this volcanic dust?
At what distance from St. Vincent were vessels covered with this dust?

uated in the island of Sumbawa, which lies east of Java, and south of Borneo.

656. The eruption occurred on the 12th of April, 1815. According to Sir Stamford Raffles, who was then governor of Java, the roar of the volcano was distinctly heard, in one direction, at Ternate, 720 miles distant from Tomboro, and in another, as far as Sumatra, at the distance of 970 miles.

657. Such vast clouds of ashes and dust were ejected, that the day at Sumbawa was as dark as the blackest night; these, rising within the sweep of the higher winds, were carried in immense quantities to Java, 300 miles distant, and hung like a pall over the island. At Macassar, 250 miles from Sumbawa, a total darkness prevailed long after the sun had risen, and volcanic dust fell an *inch* and *a half deep*. Some of the ashes were carried even as far as the island of Amboyna, which is situated 800 miles from Tomboro.

658. Near Sumbawa, such quantities of lava, cinders, and ashes fell into the sea, that they formed a cake on the surface *two feet in thickness*, and, for miles around the island, the ocean was so completely covered with this floating matter, that the progress of ships was materially impeded.

659. COSIGUINA. During the eruption of the volcano of Cosiguina, in Nicaragua, on the 20th of January, 1835, immense clouds of dust darkened the sky, and were borne by the winds to a great distance.

660. At Union, a sea-port on the western shore of the bay of Conchagua, and the nearest place to the volcano of any importance, *showers of dust* fell at intervals from the 20th to the 27th of January. It descended in the form of a fine powder like flour, and in such quantities as

---

Describe the eruption of Tomboro.
What is said respecting the ejected ashes and dust?
How did they affect the atmosphere, and how far were they carried?
What is said of the condition of the sea around Sumbawa?
Give an account of the showers of ashes and dust caused by the eruption of the volcano of Cosiguina.

to cover the earth to the depth of *five inches;* causing, for the space of *forty-three hours,* so intense a darkness that lights and torches were needed, and even these were insufficient to render objects clearly visible.

661. At Leon, the capital of Nicaragua, *showers of ashes and dust* descended on the 23d of January to the depth of *nine inches;* and at Nacaome the falling dust was mingled with coarse sand, which, together, formed a layer upon the surface of the ground *seven* or *eight inches* deep. Some of the ashes ejected during this eruption were carried even as far as Kingston, Jamaica, *seven hundred and thirty miles* distant from Cosiguina. (Art. 103.)

662. YELLOW RAINS—POLLEN-RAINS. Showers of rain, mingled with recent *vegetable* matter, consisting of the *pollen* of various plants, have been noticed for a considerable period of time. A shower of this kind once fell at Lund, in the south of Sweden, the pollen having been borne by the wind, from a forest of fir, thirty-five miles distant. A similar rain fell on the lake of Zurich, in the year 1677, and another of the same nature, at Bordeaux, in 1761. During a thunder-storm at Banff, in Scotland, on the 9th of June, 1835, a shower of *yellow rain* descended, which tinged the waters of the river Devern and of the neighboring pools with the same color. The hue in this case, as in the preceding instances, was derived from the *pollen* that was mingled with the rain.

663. A few years ago, a rain of this color extended over the western and south-western regions of the United States. At Carrollton, Ohio, the ground, after the rain, was covered with a *yellow substance;* and the same phenomenon was likewise noticed at this time at Zanesville, Cincinnati, Louisville, St. Louis, Natchez, and New Orleans. The hue of this rain is undoubtedly to be attributed to the presence of *pollen.*

664. GOSSAMER-SHOWER. A phenomenon of a very extraordinary nature was observed by the Rev. Gilbert

White, on the 21st of September, 1741, and of which an account is given in his charming "Natural History of Selborne." It appears from the statement of this gentleman, that on the day just mentioned, at about 9 o'clock in the morning, *a shower of cobwebs*, falling from very elevated regions, was observed, and which continued to descend, without any interruption, till the close of the day.

665. These webs were not single, filmy threads, but perfect *flakes* or rags—some being nearly an inch broad and five or six long, which fell with a degree of velocity that showed they were considerably heavier than the atmosphere.

On every side, as the observer turned his eyes, he might behold a continual succession of fresh flakes falling into view, and twinkling like stars as they reflected the rays of the sun. This singular shower was noticed at Bradley, Selborne, and Alresford, three places which lie in a kind of triangle, the shortest of whose sides is about eight miles long. Whether it extended farther, is not certainly known.

666. At Selborne, a gentleman, who observed this phenomenon while taking his morning ride, supposed, at first, the falling flakes to have been blown, like thistle-down, from the fields above, and imagined that he would be free from the shower when he had gained the summit of a hill that rose near his house. But upon reaching this point, 300 feet higher than his residence, he found the webs, in appearance, still as much above as before—still descending into sight in a constant succession, and twinkling brightly in the sun.

Neither before nor after was any such fall observed in these places, but on this day the *gossamer flakes* hung so thickly in the trees and hedges, that a person might have gathered them by *basketfuls*.

667. In explanation of this curious phenomenon, Mr.

Give an account of the gossamer-shower described by Rev. Gilbert White.
How does he explain this phenomenon?

White observes, that the gossamer threads, which float in the air, are the production of small spiders, that swarm in the fields in fine weather in autumn, and have the power of shooting out webs so as to render themselves buoyant and lighter than the air. If taken in the hand, they will run along the fingers, throw out a web, and sail aloft. He supposes, that, possibly, these spiders, with their webs, are carried up into the higher regions of the atmosphere by the warm and light currents of air which ascend from the earth; and that while thus elevated they have the power, perhaps, of *thickening their webs*—as some naturalists suppose—thus rendering them heavier than the atmosphere, when of course they must fall, and will thereby occasion, if they descend *simultaneously* in *large flakes* and in *great abundance*, a *gossamer-shower*.

668. Dr. Lister ascended one day, when the air was very full of gossamer, to the highest part of York Minster, and still found these filmy threads floating far above him.

CHAPTER II.

DRY FOG AND INDIAN-SUMMER HAZE.

669. DRY FOG. A peculiar haze sometimes pervades the atmosphere, which has received from meteorologists the name of *dry fog*. It is different from humid mist, for it not unfrequently prevails when no visible vapor exists in the air, and during seasons of great heat.

670. When this phenomenon occurs, the sky, although it may be perfectly free from clouds, has lost its fine azure tint, and is dull and discolored. Terrestrial objects at a distance, and of a deep color, are lost to view, and appear as if covered with a blue veil. The sun loses

---

What did Dr. Lister observe?
What is dry fog?
How does this phenomena affect the appearance of celestial and terrestrial objects?

its brilliancy, even when high in the heavens, and its light is of a reddish hue. As it approaches the horizon it assumes a blood-red color, and may be gazed at without dazzling the eyes. At times, the haze is even so thick that the solar orb ceases to be visible before it has descended below the horizon.

671. INSTANCES. In the year 1782, a remarkable fog of this kind occurred, extending over Europe from Lapland to the Mediterranean. It was succeeded the next year by another still more extraordinary. This fog, known as the dry fog of 1783, produced a great sensation throughout Europe. According to Kaemtz, its intensity was such, that in some places objects at the distance of *three miles* could not be distinguished. Sometimes they appeared blue, or else surrounded with vapor. The sun, shorn of its beams, appeared of a *fiery red*, and at noon could be looked at without injury to the naked eye. At its rising and setting it was completely obscured by the dense haze.

This dry fog first appeared at Copenhagen on the 26th of May. It reached Rochelle on the 6th of June, and was noticed, almost everywhere throughout Germany, France, and Italy, from the 16th to the 18th of this month.

It was seen at Spydberg, in Norway, on the 22d of June, at Stockholm two days after, and on the 25th it appeared at Moscow. In Syria it was observed towards the close of June, and on the 1st of July it shrouded the Altai mountains.

In England it continued from the 23d of June until the 20th of July.

During the prevalence of this phenomenon the heat was intense.

672. In the summer of the year 1834, dry fogs were noticed in various localities in Germany. Kaemtz observed one on the 29th of May, enveloping one of the peaks of the Hartz mountains.

For three days, during the latter part of this month, a haze of this kind prevailed at Munster, and the phenomenon was seen at Halle, Freïberg, and Altenberg, in Saxony, on the 28th and 29th of July.

In the northern and western parts of Germany, as well as in Holland, dry fogs very frequently occur.

673. CAUSE. The origin of this phenomenon is not yet satisfactorily explained. Many philosophers suppose it to arise, either partially or wholly, from the influence of electricity, without being able to show very clearly in what manner it is possible for this agent to produce such an effect. Others believe it to result from smoke caused by the conflagration of forests, the burning of peat-bogs, and the eruption of volcanoes. Thus Lalande attributed the *dry fog* of 1783 to electricity, Cotte to the union of metallic emanations with electricity, while other philosophers traced it to a volcanic source.

674. In the opinion of Kaemtz, the dense, *dry fog* of 1834 arose, partly from the combustion of peat, and partly from the unusual number of extensive fires that occurred in this year. While the fog was among the Hartz mountains and in the vicinity of Orleans and Basle, many peat-bogs were reduced to ashes, the fire penetrating deeply beneath the surface. One bog in particular, that of Dachau, in Bavaria, was burned to the depth of more than *eight feet*, the fire running even beneath ditches filled with water. In July there were vast conflagrations of forests and peat-bogs in Prussia, Silesia, Sweden, and Russia. The drought, which then prevailed, favored the propagation of these fires and the diffusion of the smoke.

675. The dry fogs, that occur in Holland, and in the north and west of Germany, are attributed by Finki to the combustion of peat.

676. INDIAN-SUMMER HAZE. Throughout the conti-

What is the cause of dry fogs?
What views are held respecting them?
How is the dry fog of 1834 accounted for by Kaemtz?
What is Finki's opinion regarding the origin of the dry fogs of Holland and Germany?

nent of North America, there occurs, about the close of October or the beginning of November, a warm and pleasant interval, termed the *Indian summer*, which lasts for the space of two or three weeks, and agreeably retards the approach of winter.

During this season the air is soft and bland, and a mild temperature prevails, while the atmosphere is filled with a *dense, dry haze*, that causes the distant objects of the landscape to appear as if veiled in a cloud of smoke.

677. CAUSE. This obscurity has been supposed by some writers to originate in the same way as *aqueous mists;* while others imagine it to be due to the presence of smoke, borne by the wind from the distant conflagrations of vast prairies and forests. In respect to the first view it may be remarked, that the Indian-summer haze bears little resemblance to an *aqueous mist.* It does not change into rain, and during its continuance the hygrometric state of the atmosphere is different from that which exists when moist fogs occur. The second hypothesis fails, inasmuch as it assigns a *local cause* for the solution of a *general* phenomenon—not to mention other objections which might justly be urged against it.

678. No sufficient explanation of this singular phenomenon has yet been found, but there is one circumstance connected with it which may possibly give a clue to its cause.

The Indian summer, with its genial warmth and misty veil, occurs at that period of the year when the leaves of the forest are falling, and the vegetation that covers the surface of the earth is beginning to decay. In view of this fact, the author was led to think, some years ago, that the decomposition of the decaying vegetation, which Liebig

---

What is the Indian summer?
What opinions are entertained in regard to its *haze?*
Has any adequate explanation been yet given?
What circumstance is worthy of notice in connection with this phenomenon?
In view of this fact, what has been supposed?

terms a *slow combustion,* (eremacausis), might impart that peculiar haziness to the atmosphere which is seen during the Indian summer. It was afterwards ascertained that this phenomenon was ascribed to the same cause by another observer, Dr. E. B. Haskens, of Clarksville, Tenn., who also "suggests," that the Indian-summer haze consists of carbonaceous matter or smoke produced by the oxidation of the lifeless vegetation. The warmth of this season he attributes to the same cause. These views, however, are merely speculative.

FINIS.

SHELDON & COMPANY'S
School and Collegiate Text-Books.

We would call the Especial attention of Teachers, and of all who are interested in the subject of Education, to the following valuable list of School Books:

BULLIONS' SERIES OF GRAMMARS, Etc.

A Common School Grammar.
  Being an Introduction to the Analytical and Practical
  English Grammar, ... $0 50
    This work for beginners has the same Rules, Definitions, etc., as the

Analytical and Practical English Grammar.
  A complete work for Academies and higher classes in
  Schools, containing a complete and concise system of
  Analysis, etc., ... 1 00

Progressive Exercises in Analysis and Parsing.
  . . . . 0 25

Latin Lessons, with Exercises in Parsing. . . . 1 00
  Prepared by GEORGE SPENCER, A.M., as Introd. to
  Bullions'

Principles of Latin Grammar. $1 50.

Bullions & Morris's Latin Lessons.
  For beginners, with simple lessons to be learned each
  day, and Vocabulary, etc., ... 1 00

Bullions & Morris's New Latin Grammar. . . 1 50

Latin Reader (New Edition).
  With simple progressive Exercises, references to Bullions's and Bullions and Morris's Latin Grammars, Latin Idioms, and an Improved Vocabulary, ... $1 50

Exercises in Latin Composition.
  Adapted to Bullions's Latin Grammar, ... 1 50

Key to Do. (for Teachers only), ... 0 60

Cæsar's Commentaries.
  With Notes, Vocabulary, and References to Bullions's and Bullions & Morris's Grammars, ... 1 50

Cicero's Orations.
  With Notes and References to Bullions's, Bullions & Morris's, and to Andrews & Stoddard's Grammars, ... 1 50

Sallust, ... 1 50

Latin-English and English-Latin Dictionary.
  With Synonyms, and other new features, ... 5 00

First Lessons in Greek, ... 1 00

Principles of Greek Grammar, ... 1 75

Bullions and Kendrick's New Greek Grammar, 2 00

Greek Reader.
  With Introduction on Greek Idioms, Impr. Lex., etc., . . 2 25

Cooper's Virgil.
  With valuable English Notes, ... 2 50

Long's Classical Atlas.
  Containing Fifty-two Colored Maps and Plans, and forming the most complete Atlas of the Ancient World ever published. 1 vol., 4to, ... 4 50

Baird's Classical Manual. 1 vol., 16mo, ... 0 90

Kautschmidt's English-Latin and Latin-English Dictionary. For Schools. 900 pages, 16mo, . . . 2 50

"Bullions' Analytical and English Grammar has been in constant use for several years in this and the other Public Schools of the city. It stands the test of use. The more one sees of it the better it is liked. I consider it a successful work; and I know that this opinion is shared by other masters in and out of the public service."
—JAMES A. PAGE, *Master of Dwight School.*

"We heartily concur in the above."—S. W. MASON, *Master of Eliot School.*
D. C. BROWN, *Master of Bowdoin School.*
JOSHUA BATES, *Master of Brimmer School.*

"We have used Bullions' Analytical English Grammars in our Public Schools nearly two years with success. We find them an improvement on those previously in use. Bullions' small Grammar is a fit introduction to the large one."—J. D. E. JONES, *Supt. of Schools, Worcester, Mass.*

"I have used Bullions' Analytical English Grammar some two and a half years, and am ready to give it my approval. I have not failed to pronounce it the best text-book on Grammar whenever I have had opportunity to do so. I now have a class of ninety in it, and it bears the drill of the school-room."—*Rev.* J. W. GARDNER, *Principal of New London (N. H.) Institute.*

BROCKLESBY'S ASTRONOMIES.

Brocklesby's Common School Astronomy.
12mo. 173 pages, ... $0 80
This book is a compend of

Brocklesby's Elements of Astronomy.
By JOHN BROCKLESBY, Trinity College, Hartford, Conn.
12mo. Fully illustrated. 321 pages, ... 1 75

In this admirable treatise the author has aimed to preserve the great principles and facts of the science in their integrity, and so to arrange, explain, and illustrate them that they may be clear and intelligible to the student.

"We take great pleasure in calling the attention of teachers and students to this truly excellent book. Both the plan and execution of the whole are equally admirable. It is not a milk-and-water compilation, without principles and without demonstration. It contains the elements of the *science* in their proper integrity and proportions. Its author is a learned man and a practical instructor, as the author of every school-book should be. The style is a model for a text-book, combining in a high degree, perspicuity, precision, and vivacity. In a word, it is the very best elementary work on Astronomy with which we are acquainted."—*Connecticut Common School Journal.*

"This is a compact treatise of 320 pages, containing the elements and most of the important facts of the science clearly presented and systematically arranged. It is very finely illustrated. It is worthy of a careful examination by all who wish to secure the best text-books."—*Ohio Journal of Education.*

KEETEL'S FRENCH METHOD.

A New Method of Learning the French Language.
    By JEAN GUSTAVE KEETELS, Professor of French and
    German in the Brooklyn Polytechnic Institute. 12mo., . $1 75

A Key to the New Method in French.
    By J. G. KEETELS. 1 vol. 12mo, ... 0 60

"I have examined Keetels' New Method of Learning the French Language, and find it admirably adapted for conveying a thorough knowledge of the French language. It is an easy and sure method of both writing and speaking French with accuracy and elegance."—DANIEL LYNCH, S. J., *Director of Studies in Gonzaga College, Washington.*

"The 'New Method of Learning the French Language,' by Professor Keetels, appears to be exceedingly well adapted as an introduction into the study of French. It is emphatically a practical book, and bears the mark that it has resulted from the author's own experience in teaching. I shall take pleasure in soon giving it the test of a trial in my own Institute."—OSWALD SEIDENSTICKER, *Principal of the Commercial and Classical Institute, Philadelphia.*

"I have examined several books designed for pupils studying the French language, and among them Keetels' 'New Method of the French.' The last work I consider superior to any other which I have examined, and shall use it in my classes as the best text-book upon the subject."—S. A. FARRAND, *Trenton, N. J.*

PEISSNER'S GERMAN GRAMMAR.

A Comparative English-German Grammar.
    Based on the affinity of the two languages. By Prof.
    ELIAS PEISSNER, late of the University of Munich, and
    of Union College, Schenectady. New edition. 316 pp., . $1 75

"Prof. Peissner's German Grammar has been, from its first publication, and is now, used as a text-book in this College, and has by the teachers here, as in many other Institutions, been esteemed a superior work for the end to be subserved by it, in attaining a knowledge of the elements of the German language. I cordially recommend it to the attention and use of such American Academies and Colleges as are designed to give instruction in the German language."—L. P. HICKOK, *President Union College, N. Y.*

COMSTOCK'S SERIES.

System of Natural Philosophy.
    Re-written and enlarged, including latest discoveries.
    Fully illustrated, ... $1 75

Elements of Chemistry.
    Re-written 1861, and adapted to the present state of the
    Science, ... 1 75

OLNEY'S GEOGRAPHY.

Olney's Geography and Atlas.
Revised and improved, by the addition on the Maps of the latest information and discoveries. New Plates and Woodcuts. Atlas, 28 Maps. Geography, 18mo, 304 pages, $2 40

These favorite text-books, of which more than a million have been sold, are kept up to the times by the publishers, who add the latest geographical information on the large and beautiful Maps and in the Text-Books, so as to make them worthy of the claim that they are the best works for the study of Geography now published.

PALMER'S BOOK-KEEPING.

Palmer's Practical Book-Keeping.
By JOSEPH H. PALMER, A.M., Instructor in New York Free Academy. 12mo. 167 pages, ... $1 00

Blanks to Do. (2 numbers, Journal and Ledger), each . 0 50

Key to Do. 0 10

This work is adopted by the Boards of Education of the cities of New York and Brooklyn, where it is generally used in schools and recommended by teachers. It is also recommended by accountants of prominent commercial firms, and the Press.

Whately's Elements of Logic.
By RICHARD WHATELY, D.D., Archbishop of Dublin. New revised edition, with the Author's last Additions. Large 12mo. 484 pages, ... $1 75

"This work (Elements of Logic) has long been our text-book here. The style in which you have published this new edition of so valuable a work, leaves nothing to be desired in regard of elegance and convenience."—*Prof.* DUNN, *Brown University.*

" Its merits are now too widely known to require an enumeration of them. The present American edition of it is conformed to the ninth English edition, which was revised by the author, and which contains several improvements on the former issues."—*North American Review.*

Whately's Elements of Rhetoric.
Comprising an Analysis of the Laws of Moral Evidence and of Persuasion, with Rules for Argumentative Composition and Elocution. New edition, revised by the Author. Large 12mo. 546 pages, ... $1 75

"The Elements of Rhetoric has become so much a standard work that it might seem superfluous to speak of it. In short, we should not dream of teaching a College class from any other book on Rhetoric. Communion with Whately's mind would improve any mind on earth."—*Presbyterian Quarterly Review.*

The above are the editions formerly published by JAMES MUNROE & Co., Boston, and the best in the market. They are used in all th principal Colleges and Academies in the United States.

Fitch's Mapping Plates. (In one volume, quarto.)
    Designed for Learners in Geography; being a collection of Plates prepared for Delineating Maps of the World, and Counties forming its principal subdivisions, viz., 1. The World. 2. United States. 3. North America. 4. South America. 5. A State. 6. Mexico and Guatemala. 7. Great Britain and Ireland. 8. Europe. 9. Southern Europe. 10. Germany. 11. Africa. 12. Asia. 13. Atlantic Ocean. 14. Pacific Ocean. By GEO. W. FITCH, . $0 60

NORMAL MATHEMATICAL SERIES.

Stoddard's Juvenile Mental Arithmetic.
    By JOHN F. STODDARD, A.M. For Primary Schools. 72 pp., $0 25

Stoddard's American Intellectual Arithmetic.
    An extended work, for Schools and Academies. 172 pp., . 0 50

Stoddard's Rudiments of Arithmetic.
    This work presents such parts of Arithmetic as are most useful in ordinary business computations. 192 pp., . . . 0 50

Stoddard's New Practical Arithmetic.
    Embracing methods and forms of modern business, with Analyses and many varied Examples. 192 pp., ... 1 00

Stoddard's Complete Arithmetic.
    Being in one book, the pages of the New Practical Arithmetic in the order of that book, and 120 pages of additional matter, suited for classes in High Schools. A full course in one book, ... 1 25

Key to Stoddard's Complete Arithmetic, . . . 1 00

Methods of Teaching and Key to Intellec. Arith. 0 50

Schuyler's Higher Arithmetic. (For Colleges), . . 1 25

Stoddard & Henkle's Elementary Algebra.
    By JOHN F. STODDARD, A.M., and Prof. W. D. HENKLE, 1 25

Key to Stoddard & Henkle's Element. Algebra, . $1 25

Stoddard & Henkle's University Algebra.
For High Schools, Academies, and Colleges. By JOHN F. STODDARD, A.M., and Prof. W. D. HENKLE. 528 pp., . 2 00

Key to Stoddard & Henkle's University Algebra. 2 00

"I have examined Stoddard & Henkle's University Algebra. It is a thorough and elaborate work. It combines clearness and simplicity in its method and illustrations, and constitutes a valuable addition to the mathematical works of the day."
—CYRUS NUTT, A.M., *Professor of Mathematics in the Indiana Ashbury Univ'y.*

"I have examined Stoddard's American Intellectual Arithmetic, and cheerfully recommend it to teachers and parents as a valuable elementary work, and one well adapted to the wants of pupils in the first stages of arithmetic. It is constructed upon sound and practical principles, and will be found an important addition to the text-books now in use in our Common Schools."—*Hon.* SAMUEL S. RANDALL, *Supt. of New York City Schools.*

"Stoddard's Arithmetical Series is now in general use in the schools of this county. They have stood the test for four years as the text-books in Arithmetic in our schools, and are considered by our teachers superior to any others now before the public."—*Mr.* S. A. TOURILL, *late Supt. of Public Schools of Wayne County, Pa.*

HOOKER'S PHYSIOLOGIES.

Hooker's First Book in Physiology.
For Public Schools, ... $0 90

Hooker's Human Physiology and Hygiene.
For Academies and general reading. By WORTHINGTON HOOKER, M.D., Yale College, ... 1 75

These books are text-books almost wherever they are known. The "First Book" is a text-book in the Public Schools of Boston, New York, Buffalo, and San Francisco.

"Professor Hooker's work on Physiology has been in use for the last year in the Normal School in this city, and it gives me great pleasure to express my convictions of its excellence as a text-book. In the course of my experience as a teacher, I have used the books of various authors on the subject of Physiology, but the work of Professor Hooker satisfies me much more fully than any other that I have used. It has the double advantage of being accurately scientific in its matter and arrangement, and of being expressed in correct and elegant English, a combination of the highest importance, and yet seldom attained to the extent exhibited in this book. I know of no book for which I would be willing to exchange it."—RICHARD EDWARDS. Esq., *Pres. Ill. Normal University, Bloomington, Ill.*

The Elements of Intellectual Philosophy.

By FRANCIS WAYLAND, D.D. 1 vol. 12mo, ... $1 75

This clearly written book, from the pen of a scholar of eminent ability, and who has had the largest experience in the education of the human mind, is unquestionably at the head of text-books in Intellectual Philosophy. The author's practical suggestions on the cultivation of the several faculties of the mind, aiding the student's efforts to discipline and strengthen his intellectual energies, and the numerous references to books of easy access, specifying the places where topics treated of are more fully discussed, make this book a valuable addition to the readable books of any teacher or professional man.

The Exhibition Speaker and Gymnastic Book.

Containing Farces, Dialogues, and Tableaux, with Exercises for Declamation, in Prose and Verse. Also, a Treatise on Oratory and Elocution, Hints on Dramatic Characters, Costume, Position on the Stage, Making up, etc., etc., with illustrations. Carefully compiled and arranged for School Exhibitions, by P. A. FITZGERALD. To which is added a complete System of Calisthenics and Gymnastics, with instructions for Teachers and Pupils, illustrated by numerous Engravings. 1 vol. 12mo, . . $1 25

Shaw's Outlines of English Literature.

By THOS. B. SHAW, with a sketch of American Literature, by HENRY F. TUCKERMAN, Esq. 1 vol. royal 12mo, $1 75

" Its merits I had not now for the first time to learn. I have used it for two years as a text-book, with the greatest satisfaction. It was a happy conception, admirably executed. It is all that a text-book on such a subject can or need be, comprising a judicious selection of materials, easily yet effectively wrought. The author attempts just as much as he ought to, and does well all that he attempts; and the best of the book is the *genial spirit*, the genuine love of genius and its works which thoroughly pervades it, and makes it just what you want to put into a pupil's hands."—*Prof.* J. H. RAYMOND, *University of Rochester.*

" Of 'Shaw's English Literature' I can hardly say too much in praise. I hope its adoption and use as a text-book will correspond to its great merits."—*Prof.* J. C. PICKARD, *Ill. College.*

For more full particulars, send for School Catalogue.

Address   SHELDON & COMPANY, PUBLISHERS,

500 BROADWAY, NEW YORK.

www.ingramcontent.com/pod-product-compliance
Lightning Source LLC
Chambersburg PA
CBHW031939230426
43672CB00010B/1976